Military Ethics in Professional Military Education – Revisited

Studies for Military Pedagogy, Military Science & Security Policy

Edited by H. Jung and W. Royl

Vol. 9

PETER LANG

Frankfurt am Main · Berlin · Bern · Bruxelles · New York · Oxford · Wien

Edwin R. Micewski/Hubert Annen (eds.)

Military Ethics in Professional Military Education – Revisited

PETER LANG
Europäischer Verlag der Wissenschaften

Bibliographic Information published by Die Deutsche Bibliothek
Die Deutsche Bibliothek lists this publication in the Deutsche Nationalbibliografie; detailed bibliographic data is available in the internet at <http://dnb.ddb.de>.

ISSN 1619-778X
ISBN 3-631-52864-7
US-ISBN 0-8204-7340-5

www.peterlang.de

CONTENTS

Preface

The moral and ethical implications in military leadership are on the rise. The altered face of armed conflict and the new emphasis of military missions on crisis management and peace support demonstrate the ever more crucial importance of adequate educational orientation. From an educational scientific perspective, however, not only the question of ethically appropriate behavior comes up, but also, and perhaps more importantly, the challenge of how to teach ethics and to convey immaterial contents on all levels of Professional Military Education.

Comparing the widely diverse approaches in military ethics education in the Euro-Atlantic realm in terms of contents, methodology, scope, and overall objective, a concerted review appeared appropriate to the European Forum on Military Pedagogy. Beyond that, the contributions to and the discussions of the last two annual conferences of the European Forum necessitated a stimulating revision of the entire subject in a pedagogical perspective.

Although the Cold War is long over, the field of military pedagogy is still difficult to grasp. Several attempts have been made to produce a standard definition of and orientation for military pedagogy. Definitions, however, mostly tend to be restricted to a country or a linguistic area. It also appears to be an inherent feature of pedagogy that such attempts reflect specific needs and cultural as well as local conditions. Ever since the year 2000, therefore, a regular exchange of ideas between experts in educational sciences and military pedagogy takes place on the international level of the European Forum on Military Pedagogy. This forum was founded in order to establish a platform for debate and discourse outside and beyond the boundaries of NATO/PfP as well as the domination and control of a single nation.

The attempt to reconsider the topic of military ethics in Professional Military Education undertaken here comprises mainly proceedings from the 4th International Conference of the European Forum on Military Pedagogy, held in December 2003 in Vienna, Austria, under the title *"Military Ethics as an Educational Challenge"*. This volume also incorporates contributions presented at the 3rd Conference of the European Forum, which was held in Zurich in September 2002 and covered topics around military pedagogical research, adult education and the role of military pedagogy in coping with future tasks.

This book is being both edited and published on a joint basis by the two institutions responsible for conducting the two above mentioned international conferences on military pedagogy, the Institute for Human and Social Sciences of the National Defense Academy, Vienna, and the Department of Military Psychology and Military Pedagogy of the Military Academy at the ETH Zurich.

This compilation advances the subject area in three major chapters. The first tackles challenges as they occur in teaching and studying ethics, from a philosophical and scientific perspective. The section outlines the philosophical agenda for the ethical debate in military education and offers ideas on how to relate military pedagogy to the demands of civil-military relations. Ethical dimensions in teaching and learning are examined with a particular view on how that plays out in the information age. This issue is further scrutinized with regard to the task of teaching values and professional ideals. The section concludes with deliberations on the impact of social and cognitive skills on the selection of military officers and psychological implications in teaching ethics.

The second chapter turns to the application of ethics in particular human contexts and operational settings. The topics covered range from the specifics of adult education to operational thinking and network-centric warfare, a psychological account of how to create sensibility for applied ethics, and action research, to some concrete examples of how morals and ethics are taught at certain armed forces institutions throughout the world.

The third chapter is dedicated to ethical challenges as they arise from asymmetric conflict and warfare. It deals on the basis of political and military implications with ethical questions regarding asymmetric fighting but, most notably, looks at the moral challenges arising for regular military forces when fighting asymmetric adversaries and deliberates on the subject of preparing for asymmetry by analyzing Joint Vision documents for armed forces' development.

First and foremost, the editors would like to thank Prof. Dr. Wolfgang Royl, Professor emeritus of the Bundeswehr University Munich, and Dr. Hermann Jung, from the Austrian Ministry of Defense, for making the publication of this volume possible within the series on Studies of Defense Pedagogy, Military Science, and Security Politics. This book represents the ninth publication issued in this series which was launched back in 1989, at a time when the beginning change in the security environment should usher defense establishments into a new era of manifold challenges.

The gratitude of the editors also reaches out to all the authors who contributed to this volume. The articles and essays build a basis for substantial reflection and valuable academic dialogue, which should be continued in the future. It has to be emphasized, however, that the opinions expressed in this book solely represent the personal views of the authors and do not represent any official policies.

In addition, the editors would like to name all the others who helped to make this publication possible: the colleagues in the Institute for Human and Social Sciences, Dr. Guenther Fleck, Andrea Grasser and Eva Nemec, for collecting the contributions and executing the lay-out, the visiting research associate Mr. Samuel Schubert for his supporting advise, and US Air Force Major Johnny

Barnes for his editing assistance to the first editor during the latter's teaching assignment at the Department for National Security Affairs of the Naval Postgraduate School in Monterey, California.

It is the consensual intention of the editors and contributors to the European Forum on Military Pedagogy to continue the joint efforts in the field of military ethics and Professional Military Education in the future.

Edwin R. Micewski & Hubert Annen

Monterey and Zurich, September 2004

Philosophical and Scientific Challenges in Teaching Ethics

Edwin R. MICEWSKI, National Defense Academy, Vienna

ON THE PHILOSOPHICAL FRAMEWORK FOR THE ETHICAL DEBATE IN PROFESSIONAL MILITARY EDUCATION

When thinking about the position of men and women in military uniform, particularly of leadership personnel, one becomes aware of how vast the scope of both intellectual and moral difficulties officers have to face in their duty is. It is exactly the significant difference to other occupational worlds which gives us an infallible idea as to the necessity of comprehensive education for military leaders. The aptitude and, if necessary, readiness to apply force and thus put one's own and the life of others at risk, necessitates personalities who are actually and in the true sense of the word able to take on responsibility, not only for themselves but, above all, for others.

The philosophical framework for the ethical debate in military education is almost as comprehensive as philosophy itself. An all-inclusive understanding of education, aimed at enabling somebody to mold himself intellectually on the basis of autonomy and self-determination, should grant any human individual knowledge and wisdom built upon the grand traditions of human efforts in the fields of humanities and culture. In this perception, education reaches out far beyond the immediate knowledge and capacity one might need to carry out his occupational tasks and, thus, embraces all that unconfined and freely available knowledge that results in holistic competence and a profound ability to cope with responsibility.

Against this comprehensive backdrop of leadership responsibility, the philosophical framework for military professional education is constituted by the systematic of the philosophical disciplines. It thus comprises the major fields of both theoretical and practical philosophy, ranging from epistemological implications, logic, metaphysics, and aesthetics to the subject areas of moral philosophy, ethics, and political philosophy.

An all-embracing philosophy of the military, however, operates in two pragmatic respects: The intellectual setting is, on the one hand, provided by the distinctive nature and the organizational peculiarities of the military world and, on the other hand, is being built by the social and political conditions of open democratic societies into which both soldier and military organization have to integrate themselves.

Educational Perspectives

In the comprehensive understanding as outlined here education comprises the dimensions of basic *knowledge*, basic *capabilities*, and basic *virtues*. Hence, education also has to encompass the realm of basic morality and can therefore never be thought devoid of the idea of freedom. Both the end and purpose of education must always be seen in the attainment of a certain degree of personal autonomy.

This is very much in accordance with the enlightened dictum of Immanuel Kant, who called upon the human to release himself *"from his self-incurred tutelage."* Now, one is supposed to add that this tutelage might well be imposed 'otherwise'; nevertheless, Kant is touching upon a decisive aspect when he cites laziness and cowardice as the major reasons that keep the majority of people from making use of themselves without the guidance of anybody else. It can be convenient to remain in the stage of tutelage because *"if I have a book, which understands for me, a pastor who has conscience for me, a physician who decides my diet, and so forth, I need not trouble myself."* (Kant, 1963: 3 [35]) However, there is no doubt that quite a few might thoroughly enjoy having such dependent individuals as coworkers or subordinates.

This preeminent claim to education posed here has somehow gotten lost in the process of the explosion of knowledge that we witnessed in the last decades and which led to the ever more illimitable exhibition of information in modern societies. No longer was the majority and emancipation of the individual in the foreground, but rather his regulation aiming at teleological handiness and functionalism. Education became more and more a kind of mere 'management of knowledge,' mostly understood as the accumulation of facts and data, at best available on line and easily recallable from a network. (Liessmann, 1999: 13)

The consequences of this purely instrumental and teleological understanding of education turned out to be problematic, as a generally deficiently oriented and insufficiently self-reliant human was being produced. This is why we recently observed the turn to a concept of education which emphasizes the idea of a harmoniously cultured human as well as the ideal of a completely enlightened person. Thereby the concept of education is raised up to the level of a cultural task, if not duty, of ethical dimension; a claim that calls upon the individual as well as any political system.

In a social and political respect, education becomes, at the same time, a precondition for the formation of a democratic community, in particular under the conditions of increasing complexity. All members of a political community will only be able to entirely participate in the formation of the political and social order when they are culturally educated. A form of education, which - not constrained by limitation to occupational knowledge and efficiency - encompasses also the manifold modes of social competence, including the aptitude to partake in a conscientious manner in relevant processes of political formation. Civilian supremacy upon which the military operates notwithstanding, this 'emancipated' *loyalty*[1] should not only be expected from any citizen, but also from every soldier.

As the soldier belongs, like any other occupationally engaged person to both an existential and a professional environment, it is just a matter of logical inference to assume that education dissociates into *general education* on the one hand, and *occupational education* on the other.

Practical wisdom, which includes the value-related interpretation of existence, i.e. an orientation toward the world as well as existential values, joins hands with a technical-instrumental education which, together, should enable somebody to *"put himself as well as his relation to the world in good order"* (Ritter, 1971: 921).

Political-Philosophical Perspectives

Before looking at the professional world of the soldier, one has to bear in mind that armed forces find their position and societal integration in a comprehensive political context only. In the first place military organizations are primarily bearing political character. As an instrument of politics, armed forces confer - under the presupposition of the primacy of policy and the willingness of political leadership to make use of this tool - any policy with the capacity for self-determination, protection of peace and security, and international cooperation and solidarity.

This particular relationship - the art of politics precedes the art of war - has been particularly posited by the Prussian theorist of war, Carl von Clausewitz, who had coined this so frequently misinterpreted and misunderstood dictum, nonetheless familiar quotation, that *"War is the continuation of policy by other means."* (Clausewitz, 1976: 76) This sentence hints - in modern parlance - among other things, at the fact that the political factors are cardinal to the mere military ones.

In an educational sense, this dependency of the 'guardians' on the political sphere has already been acknowledged by the ancient philosopher Plato. He stipulated that the guardian has to demonstrate a philosophical inclination to acquire knowledge (without intending to become a philosopher himself), for education alone enables the guardian to come up with a friend-foe discrimination and protects him from being politically misused. (Plato, Politeia, 376b-c)

Within the political system, the professional culture of the military rests - as is the case with any other occupation - with the purpose of the military

---

[1] Loyalty is taken here in the sense of an individual's readiness to take on justified claims stipulated in public interest or need.

organization and the tasks it is supposed to fulfill for the political system. In generalizing conciseness, the suitability of the military lies in the aptitude of putting through the interests of the nation under the exceptional circumstances of armed conflict and national distress. This allows a conclusion on the nature of the military which has to be seen in the ability to legitimately threaten with or apply the use of force. From this capability to successfully endure the conditions of armed conflict, then derive the requirements of military education and training, which lead to the peculiarities of the military world becoming accustomed to physical and psychic strain, discipline, hierarchical structures, strict relations between command and obedience, and all those other features that help to prevent the cohesion of any part of the military establishment from breaking up under the most challenging and strenuous conditions of armed conflict.

The professional world of the military man (woman) is part of the existential environment in which the soldier finds himself in a twofold fashion, as both a citizen of the society and an executive organ of the state. The framework to these professional and existential conditions is being built by both the political and societal factors and conditions into which military and soldier are supposed to integrate themselves. The more or less timeless characteristics of the military profession as they were outlined above have to be joined now with the dynamically and constantly changing strategic circumstances of the security environment. In conjunction, this results in the profiles of roles, tasks, and missions the armed forces have to take on in the national, but ever more importantly, in the international arena.

Modern military professionalism, in the three-dimensional context of military, society, and politics, has to meet the challenges as they derive from both the *functional* and *societal* imperative of security. (Huntington, 1957: 2ff)[2] In other words: Armed Forces have to cope with instrumental as well as intellectual challenges and have to be able to act on both a *material* and a *sensual* level. This constellation is exemplified in the graphic below:

Position of Armed Forces in Society and State

Functional Imperative	***Societal Imperative***
Security Environment Threat Scenario (objective - official) Task Assignment to Armed Forces Implementation of Means	Values and Attitudes Power Relations within the Political System Interests (individual-collective) Threat Perception (subjective - private) Powers of Social Formation (Ideologies)

[2] Huntington speaks of the *functional imperative* and the *social imperative* which, in their unity, determine tasks and roles as well as position and acceptance of any armed force in their political environment.

(Inter) National Strategic Conditions (General Parameters)	*Social & Societal Conditions* (General Parameters)
Significant Alteration of Strategic Environment Massive Reduction of Land Forces Reduced Significance of National Defense Multi-dimensionality of Security ("Comprehensive Security") Increased Significance of International Crisis Management and Peace Support	Individualization (Egocentrism) Professionalization / Specialization Cognitive Mobilization: Increased Importance of Communication and Information Military Security Considered Less Crucial (Old conceptions of security doubted) Armed Forces / Conscription Lose Acceptance
Functionality "*Material* Level"	***Legitimacy*** "*Mental* Level"
Stability & Acceptance	

The societal imperative is predominantly of sensual nature, meaning that it refers to 'soft' dimensions of security such as attitudes and values, rather than to 'hard' facts of security like threat scenarios or power potentials, which are generally easier to assess and to measure.

The functional imperative of security has to be seen in the changed face of armed conflict and the ensuing new challenges the armed forces have to meet in the framework of an extended, more comprehensive understanding of security, running parallel with the increasing internationalization of security policies. The societal imperative arises from the conditions of modern societies with their increasing tendency for individualization - cognitive mobilization, as Lawrence Kohlberg put it - and the reduced need for immediate national defense and the declining acceptance of hitherto defense concepts and defense roles. This brings the demand for comprehensive education of the soldier to the fore of educational efforts. Proof of this is given by the numerous projects dedicated to the reformation of Professional Military Education (PME) and training, as they are to be recognized in more or less all nations affected by the drastically changed world since 1989.[3]

[3] Within the framework of the Partnership for Peace Initiative, for example, a Consortium of Defense Academies and Security Studies Institutes was implemented in 1998, aiming to coordinate and foster joint research, education, and training with respect to security and military affairs. Hitherto the consortium has established seventeen study and working groups to deal in different fields with educational matters.

Inferences for Professional Military Education

The changes in the security arena have also drastically altered the tasks and missions carried out by the armed forces. Along with the emergence of the concept of comprehensive security, a considerable amount of initially non-traditional tasks be they of humanitarian, ecological, human rights, or social nature came into the grasp of military organizations. This is due to the fact that those tasks can only be dealt with by, or by assistance of, institutions and organizations that have the required organizational structure, the human and material resources, appropriate means of communications, and the ability to both act and react rapidly.

Military forces will, thus, play a significant role in the 'new' security policy as they are inevitable in coping with the manifold categories of strategic and operative security measures within the framework of crisis and conflict prevention and general peace support. Certainly, while the armed forces become an inexorable stabilizing element of conjoint, supranational politics, they, by and large, lose their classical function as being a mere maintaining factor of national power politics. In this respect, analysts speak about the constabularization of armed forces which, though only to some degree, makes sense. (Janowitz, 1971: 417) This goes too far, however, when attempts are being made to ignore the difference between peacetime and wartime and to draw a picture of the soldier as being just a martially looking policeman. The motives, which might drive such approaches notwithstanding, military forces need to be adept at carrying out traditional combat missions although those missions might no longer remain in the foreground of armed forces task profiles.

Particularly for the armed forces of European countries who have not seen high intensity conflict or even outright war for decades, this new development offers risks but also great chances. Above all, soldier and armed forces stand a good chance to found a new identity that turns away from the mere functionality of the threat and, if need be, the application of force to the image of an instrument, which contributes efficiently to a stable and vital subsistence of not only one's own countrymen but the whole of mankind. (Däniker, 1992: 165-188)[4]

From all this we have to deduce a certain *"civilianization"* of armed forces that entails a demand for increased civil competence, particularly on the level of commanders in respect to social, ethnic, historical, in a word, cultural influences on crisis and conflict scenarios. To think in this humanistic context, i.e. of the respect for life, protection of culture and cultural property, the ability to cooperate with soldiers who stem from different cultural, linguistic, and political systems, and the competence to deal with humans in crises zones that lie outside the own political or cultural area are required.

---

[4] In particular his Seven Theses.

Additionally in this context, we must not forget about the increased demand to legitimize both the existence and acting of the power instruments of nations and alliances in the framework of contemporary challenges. Moreover, the kernel of soldiering, which even under the altered conditions of security has to be seen in the application of physical force, needs to be intellectually argued and justified. The soldier, and particularly the military officer, has to be able to credibly demonstrate the true essence of military duty in the context of the modern, tendentiously pacifist societal consciousness that basically aims at the peaceful and non-violent resolution of conflicts.

Thus, the scope of both intellectual and moral difficulties that have to be covered by military education and formation will be fairly vast. It commences with the necessity to fathom the correlation between state, law, and violence and force as well as both the moral-ethical and political foundation and legitimization of the soldier's profession. It ends with the abilities, which are demanded by the new technological challenges and the nature of the tasks the armed forces have to fulfill in an increasingly international and multi-cultural framework. Briefly stated: A broad array of cultural scientific knowledge and understanding will be required on all levels of military duty in order to prevail successfully in both missions as well as in the critical environment of modern societies.[5]

As one example might serve here the educational reform process that the armed forces of the United States have already embarked upon in the 1980s. Military leaders should be enabled to adapt all the relevant political, strategic, and operative dimensions into the formulation of military strategies and to unite the totality of political, military, and civil factors that unfold in a given mission area. A scholarly expression of this is found in the curriculum of the US Army War College:

> *"The college curriculum includes not only military problems, but also studies of the social, economic, psychological and political factors which affect our national and foreign policy formulations, strategic overseas areas, and a consideration of many new developments in the fields of science and technology. It provides a place where mature officers of all services and their civilian counterparts can come to*

---

[5] To determine appropriate educational contents for every level of military formation and education would go far beyond the purview of this paper. It has to be dealt with in consortiums and expert groups as mentioned above in reference to the PfP initiative. In both national and international forums, the necessary contents need to be defined and accorded as to make leadership personnel fit for the multi-facetted challenges of the 21st century. This will require close civil-military cooperation not only in the mission fields of Peace Support Operations, but also and already on the level of joint education; first and foremost to mention the closer cooperation between civil and military institutions of security studies and academic education.

grips with the problems which face our nation. Here, emphasis is paced on freedom of thought and action. Qualitative learning is considered more desirable than quantitative." (awcci@carlisle-emh2.army.mil)

A Perspective on the Future

At all times and in all armed forces there has constantly existed anxiety to come in contact with education in the sense of higher intellectual demands. This negative attitude sprang mostly from the fear that too much intellectuality could lead to selective, critical, and even uncontrollable behavior. Although generally open toward technological innovation, the resistance against cultural alterations could be observed all along the way. Anti-intellectual attitudes manifest themselves, i.e. in steps to respond to intellectual challenges by mere technocratic and instrumental answers (standard procedures, rules) or to reduce those challenges to seemingly technically resolvable ones.

There is much evidence throughout the military forces in the Western world that the educational deficits of earlier times will be overcome. Efforts are being made to bridge the chasm between the intellectual-educational requirements of the new security environment as well as the challenges today's societies pose upon the military leader on the one hand, and the actual reality of cultural knowledge and competence on the part of military personnel, on the other.

However, if one agrees with the fact that any military officer - in an appropriate understanding of the concept - should belong to the intellectual 'elite' of a country's society, there still is a long way to go.[6] As one decisive precondition for the successful pursuance of such a course, a fundamental change of consciousness toward the significance of education in this more sophisticated sense outlined here will have to take place across all the ranks in the military. Any conception of education that aims at making the military leader fit for the challenges of the 21st Century will have to take into consideration that not every soldier and officer shall necessarily be a 'military intellectual,' but in any event, shall have to represent an 'intellectual military' as well. (Janowitz, 1971: 431-435)

With this helpful differentiation, Morris Janowitz, the renowned military sociologist introduced the idea that for every soldier in a leadership position it is obligatory to integrate a certain *'intellectual'* dimension into his professional service. The 'military intellectual,' however, as a scholarly and scientific type of

---

6 The concept 'elite' (selected minority, leadership stratum) is taken in the sense of individuals in modern pluralistic societies who, because of their both human (intellectual and moral) and professional qualities, constitute leadership groups.

character who is not considered to take on immediate military leadership tasks, does assist leadership personnel to achieve and maintain that status.

Therefore, the relationship between the intellectual militaries and the military (or other) intellectuals should ideally be a complementary one, characterized by mutual understanding and shared appreciation. Achieving this should also be one particular goal of comprehensive professional military education.

References

Kant, Immanuel: What is Enlightenment? In: Beck, L.W. (ed.) (1963): On History. Macmillan Publishing Company, New York.

Liessmann, Konrad Paul: Wissen ist Macht, offenbart aber auch Ohnmacht. In: Der Standard (1999, April 7), Vienna.

Ritter, Joachim (ed.) (1971): Historisches Woerterbuch der Philosophie. Band 1. Basel/Stuttgart, (this section transl. by E. M.).

Clausewitz, Carl von (1976): On War. Princeton University Press, New Jersey, Book One.

Plato: Politeia. 376b-c.

Huntington, Samuel (1957): The Soldier and the State. The Theory and Politics of Civil-Military Relations. London.

Janowitz, Morris (1971): The Professional Soldier. A Social and Political Portrait. Macmillan Publishing Co., New York.

Däniker, Gustav (1992): Wende Golfkrieg. Vom Wesen und Gebrauch zukünftiger Streitkräfte. Frankfurt a. M.

USAWC, Carlisle Barracks, PA 17013, AWCCI@carlisle-emh2.army.mil .

Wolfgang ROYL, University of the Bundeswehr, Munich

MILITARY PEDAGOGY AND THE CONCEPT OF LEADERSHIP DEVELOPMENT AND CIVIC EDUCATION

In this essay, research questions are legitimized by the institutional reference that characterizes both military pedagogy and the concept of Leadership Development and Civic Education (Innere Führung). Military pedagogy is defined as a partial discipline of educational science and is thus understood to be universalistic, just like educational science itself. On the other hand, Leadership Development and Civic Education (LDCE) is regarded as the doctrine of leadership, training, and education of the armed forces in the Federal Republic of Germany. The question is raised how such a doctrine can be positioned within the field of military pedagogy, which is focused on the pedagogical activities in armed forces in general. This scientific focus on knowledge is global in terms of the description, comparison, analysis, and evaluation of different national military pedagogies. It is universalistic with respect to the reconstruction of soldiership to secure peace in freedom.

In contrast to military psychology and military sociology, the topics of military pedagogy are the drafting, implementation, evaluation, and dissemination of curricula concerning leadership, training, and education of the armed forces. The central idea of military pedagogy with respect to the learning goals to be achieved step by step is gained from the reconstruction of the soldierly image of the individual, whose ultimate core statement is the individual preparedness to risk one's life for public welfare.

Soldiers killed in action remind the surviving world to pursue in a more intense way the reimplementation of a peaceful way of the living together of all people. The death of soldiers in battle carries a symbolic value expressed in the form of commemoration days, plaques, and monuments, which point to the fact that within the individual transcendence of the soldier killed, the antagonism between attacker and defendant becomes obsolete. At the same time, it is the task of society to care for war-disabled persons. These obligations are anchored in the public consciousness through the interaction of social ethics and military pedagogy.

A basic conclusion that can be drawn from the analyses that have been conducted in the field of military science so far and that has thus retained its future-oriented quality is: It is not only the quality of armament which is vital for the defensive capability of an army; the personnel's willingness to defend their country is also crucial.

This means that armed forces should not only focus on the training of personnel in terms of carrying out defensive actions but also on their willingness to do so. While there are countries and state-like groups which wage wars of aggression or which violate the peace by their war-like activities, the armed forces' role in terms of security policy is considered to be the defense of their own territory, of human rights, the deterring of terrorists, as well as other, undefined military threats. In order to prepare enlisted personnel and officers for these tasks they need to know how soldier-adequate behavior is described. If general ideas of what a soldier's behavior should be like are developed into typified concepts, the military image of the individual takes shape and, with regard to the Western world, leads to the concept of soldiership integrated into democracy. In other words: What is the scope of commitment in terms of public welfare, loyalty towards the state, and recognition of the reason of state? What selfless obligation must a conscript or career soldier have toward the best interest of the democratic community? What kind of status does society accord to the image of an individual who makes use of armed force and is willing to sacrifice his life?

Military pedagogy tries to find answers to these questions and deals with military everyday culture and military ethics, i.e. the system of military virtues and how and to what degree they can be imparted. This scientific discipline reflects what scholars like Huntington (1957), Janowitz (1960), and Moskos (1977), among others, contributed to the professionalization of the military trade already decades ago. Their findings are still important to the military doctrine referred to as Leadership Development and Civic Education, which was especially developed by the Bundeswehr as an educational model in the field of military pedagogy.

The following distinction between educational science and military pedagogy is made for reasons of scientific systematology. While educational science reconstructs different forms of pedagogy, it is the task of pedagogy to reconstruct the image of the individual (Paschen, 1979). For example, issues like the image of officers or soldiership within democracy are topics of military pedagogy. It depends on the reconstruction of military pedagogy through analyses in the field of educational science to determine to what extent military pedagogy has the right to exist. Such an analysis could serve to establish whether the armed forces need their own institutional pedagogy, i.e. military pedagogy, or whether established branches of educational science such as pedagogy for job training or adult education cover the fields of leadership, training, and education for the armed forces adequately not only in theory but also in practice, both in research and teaching.

This essay focuses on the relationship between military pedagogy and Leadership Development and Civic Education. First, the fields of application of military pedagogy and Leadership Development and Civic Education will be defined. Then, a text comparison will make a statement on the relationship between these fields; and finally possible conclusions will be drawn.

The Application of Military Pedagogy

Pedagogical activities focus on developing personalities. Pedagogues support this development by a specific combination of learning processes which contribute to the student's education and training. They counteract tendencies that have a negative influence on the image of the individual which is characterized by moral independence. Whether such an image can be achieved through pedagogy depends on the reconstruction criteria of educational science, the criterion of *"original sense,"* and the criterion of contra-facticity. (Paschen, 1979) With regard to military pedagogy and put into an interrogative sentence: What is the original sense of military pedagogy? This question focuses on the central idea of military pedagogy. And, concerning contra-facticity: Should military pedagogy be the way it is right now?

Military pedagogy focuses on that part of education which represents the development of the individual's and the force's military competence as to taking action. With regard to the theory of personality, both the individual's and the force's competence can be shaped as a role segment. As a result, citizen soldiers, if, for instance, involved in a debate about UN missions, are able to justify the purpose of such operations from a soldier's point of view. The military training they received enables them to gather background information. When dispatched to take part in a UN mission, it is military action competence that enables personnel to act in a professional way in support of the mission.

Pedagogical activities become more sophisticated if the soldier's role behavior is developed not only in a peripheral way, but if some kind of psychosocial identification with soldiership takes place. This is particularly related to military officers. If being an officer means more than just following a career path or merely adopting a specific role behavior but becomes part of one's own personality and identity, these officers would not just say: *"I am an officer"* but *"I want to be an officer."* Any such statement putting emphasis on one's own image would prove that an officer has managed to combine professional role behavior and personal identity. It is by no means all that easy for soldiers to identify themselves with their profession, as in Germany we live in a society where sometimes military ceremonies held in public have to be protected by police forces and where law courts debate at great length whether or not soldiers may be called murderers. Even those who have joined the Bundeswehr identifying themselves fully with the ethics of their profession may once in a while feel personally offended.

It would be to the benefit of any army if counseling were provided within the framework of the personnel management system to counter anything that might be detrimental to the soldiers' professional identification. There are officer-training candidates who have been accepted as career officers but refrain from taking up any such career because they have realized that the personnel manager does not note their outstanding achievements as students and soldiers as they - quite rightly - expected him to do. Officers who file complaints against superior officers sometimes find that it is not so much the content of their complaints that is looked into but rather the legal aspects of the case.

The field of military pedagogy should thus not only concentrate on identifying the impacts of personnel management in the military but also give sound statements about its positive effects. Building on that, a doctrine centered on military pedagogy would then have to ensure such positive effects. With respect to the German Bundeswehr it is the military doctrine of *Civic Education and Leadership* that reflects such a pedagogic approach.

Some of the senior officers think that what they are already taking into account in their daily duties is exactly what has been described above, albeit in a more complicated way. At the same time, the same officers would not hesitate to complain about the vanished 'esprit de corps' among officers, the lack of acceptance in our society regarding the military profession, and the steady, and thus alarming, increase in the number of conscientious objectors. This problem is generic and merely shifted when transforming a conscript army into an all-volunteer force.

How can we rebuild the image of the armed forces in a democracy to an appropriate extent? Can we draw any lessons from the events that took place nearly 200 years ago?

The Army Reform of 1806 would not have had such a positive outcome had it not been for the national uprising, to which philosophers like Johann Gottlieb Fichte, who delivered his *"Addresses to the German Nation,"* contributed. Or think of Ludwig Jahn, who revived the nation's confidence by instilling in people an enthusiasm for sports, and Schleiermacher, who characterized the state as the guarantor for collective morals. The times of national uprisings are over, but being prepared to defend peace in freedom is still crucial.

My thesis is that we should try to establish a theory concerning a universalistic security policy and a teaching model thereof, as well as to make educational institutions - national universities and colleges – to participate in this endeavor.

The attribute *"universalistic"* is connected to the phenomenon of globalization and its structure. The attacks on the World Trade Center on September 11th, 2001 in New York clearly showed the threats we are facing.

The purpose of universalistic security policy is to not only focus on peace efforts in regional trouble spots but to pursue them on a worldwide basis. While there is a UN Charter created as an international document demanding a peaceful coexistence of nations, what is missing are the dispositional traits that cause a certain behavior with some degree of probability and permanence. Intentional, i.e. planned behavior depends on normative beliefs and moral obligations, on personal beliefs, attitudes, subjective norms, and the results of behavioral control. The *"Theory of Planned Behavior,"* which should be introduced at this point, is a good theory to categorize the field of application of military pedagogy. (Ajzen & Madden, 186: 453-474)

Theory of Planned Behavior

We have assumed that the field of military pedagogy is a branch of educational science dealing with research and the testing of means and methods by which norms and values taught could influence the behavior of soldiers. Schiller pointed to the subject-object relation included therein: *"Only if prepared to risk thy life, thou will be larger than life."*[7] The chart of path analysis (Fig.1) explains the above mentioned theory, the usefulness of which has been proven in psychological studies of different topics. This model is supposed to show the different pedagogical efforts needed to pave the way for planned behavior. From an educational point of view, the planning of behavior is reflected in learning objectives and educational goals. With regard to military pedagogy, the planning of behavior has to take into account the variables that have been characterized as internal personality criteria in the behavior model.

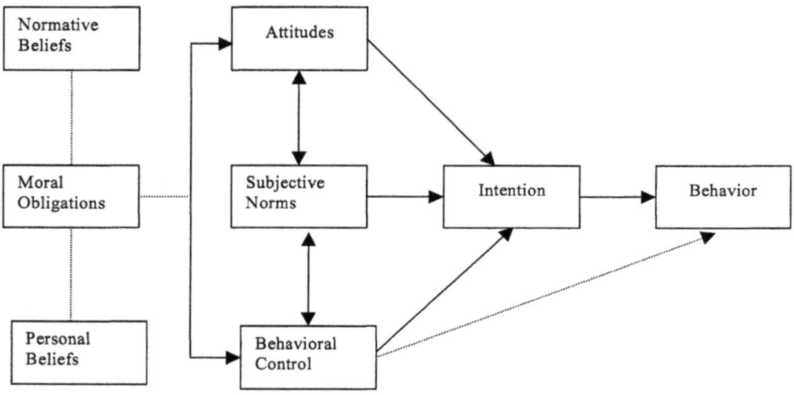

Fig. 1: Path Analysis for Planned Behavior

---

[7] In: *Wallensteins Lager.*

Concerning the field of application of military pedagogy it would be necessary to investigate the normative beliefs of officers who identify themselves with their profession, to what extent they have managed to adopt normative beliefs as their personal ones, and how they assess these beliefs from an emotional perspective. The variable *"attitudes"* comprises a person's affective and evaluative reactions when translating personal beliefs and intent into overt behavior. Subjective norms are other people's expectations which we have internalized. *"Behavioral control"* refers to the internal checking of the necessary requirements for somebody to muster the confidence to put planned behavior into action.

It becomes evident that the theory of planned behavior does not explain spontaneous individual behavior but behavior which has been worked at in the armed forces in the long term through leadership, training, and education. As a result, the field of application of military pedagogy is restricted to the institution of the armed forces to which, independent of national borders, it generally applies.

If, from an educational science point of view, it is the task of pedagogy to reconstruct the image of the individual, military pedagogy should find answers to the question of what the image of citizen soldiers should be like. In other words: military pedagogy deals with soldiership integrated into democracy. The reconstruction process is caused by behavioral deficits in the military sphere. Just an example to clarify this point: An officer in the rank of rear admiral coming to the conclusion that the Bundeswehr is a neurotic army (Schmähling, 1991) without doubt makes a statement about a deficit. Military pedagogues would not get on this train of thought straight away but ask how this officer could make this diagnosis. If it were correct, the next step would be to study how such a deficit could be overcome. If, however, the rear admiral's diagnosis was wrong and just made to compensate his frustration about his career through disloyalty, it would still be necessary to see whether the selection process which paved the way to such a high military rank was deficient, particularly bearing in mind all the performance assessments which were made during his different assignments as a navy officer. The assessment report used by the Bundeswehr requires the rating officer to provide information about *"outstanding personal characteristics, comradeship, professional self-perception, achievements during mission and additional information."*[8]

Military pedagogy designed as biographical research would try to find out about the ideas entertained by the supporters of the Prussian Army Reform of 1806, Scharnhorst, Gneisenau, and Clausewitz. In this respect, it would be particularly interested to investigate if officers' loyalty to the king and the army depended on

---

[8] Section H / form no. 0119/98/V.

whether their promotions met with their expectations. Military Pedagogy would also have to compare the quality of the texts which were written back then and are written today on the restructuring of the army. Furthermore, it would be legitimate to ask what conditions would have to be preserved or created in order to pave the way for the development of a leadership personality.

Within the framework of comparative military pedagogy it would be necessary to raise the question about differences in the armed forces of various countries in the fields of leadership, training, and education of soldiers. The importance is particularly evident when these countries are members of a military alliance which plans for and engages in multinational missions. In order to deal in a more concrete way with this problem of military pedagogy as far as mental cohesion in multinational units is concerned, *"Euro-military identity"* was defined as a developmental task. (Royl, 1992) Instructional objectives which meet this requirement are supposed to illustrate the points made:

- Name historical events where European countries jointly countered military threats (*knowledge*).
- Explain why the Oranian Army Reform plays an important role in the history of military pedagogy (*understanding*).
- Find examples which show how the Prussian Army Reform has been put into practice regarding military pedagogy (*application*).
- Make an assumption as to the conditions under which a combined supreme command structure proves to be efficient in military terms (*analysis*).
- Give a brief statement in your own words on the factors that unite European states in terms of security policy (*synthesis*).
- Assess what benefits a knowledge of European military history would have for military pedagogy to promote cohesion in multinational units (*evaluation*).

The thesis goes that it would be important to create group cohesion through mutual understanding of national cultural issues and shared values among future personnel of the *"United States of Europe"* and, in anticipation thereof, for multinational units such as the Eurocorps, the Franco-German Brigade, the German-Dutch Corps and the Bundeswehr Special Forces Command.

In a request to the EU Commission concerning the TSER[9] program these ideas have been put into the concrete form of a research project which serves as an example for further research activities. Based on the guiding image of *"Cohesion through Acculturation,"* the title of the project was "Curriculum

[9] TSER = Targeted Socio Economic Research.

Development for NCOs and Officers in Multinational Military Units to promote *Euro-military identity"* as a basis for Defense Readiness in a European and Global Dimension."

It is intended to initiate a network cooperation in order to promote research studies, which stress the cohesion of NCOs and officers in multi-national military units, especially in the Eurocorps, the Franco-German Brigade and multinational UN missions. The intended studies aim at two particular aspects: (1) Development of curriculum materials for mental training in European culture; and (2) Implementation of curriculum modules in military education and evaluation of its impact by the aptitude-treatment interaction design.

The first step in the research agenda would be to find out what is general in the comprehension of security policy, the cultural background and professional convictions between officers and NCOs in military forces of different national origins, followed by analyzing the cultural significance of the esprit de corps of supervisors belonging to different national contingents within multinational units. The third step would then be to look at the appreciated curriculum modules containing elements of European culture by each nation involved. Fourth, military training institutions ready to allow the implementation and evaluation of training in the development of Euro-military identity should be recognized in order to finally, fifth, apply didactical and methodical arrangements in related education. Evoking an esprit de corps in multinational units will be offered, among others, by the Ecole Militaire Superieure (Zurich), the Leadership Development and Civic Education Center of the Germany Federal Armed Forces (Koblenz), the Netherland Royal Military Academy (Breda), and by the Centre D'Etudes En Sciences Sociales De La Défense (Paris).

In comparative studies, sociologists specialized in the military field would consider the army a social institution, connect the recruitment of the officers corps to social structure and make an analysis of social processes which occur within a military unit. A study that investigates the number of suicides in an army and the reasons for soldiers taking their own lives would be considered to be part of military sociology. (Preuschoff, 1988) In order to recognize suicidal tendencies, military psychologists would carry out test studies to gain diagnostic data about soldiers' personality structures. Military pedagogues would examine how important education theory actually is for the armed forces and how important it should be. They would apply means and methods to convey the sense of military service as a stabilizer for one's personality and, if appropriate, they would put subjects on the curriculum which serve in a positive way to define one's own position within the cultural universe.

The aim of military pedagogical activities consists in teaching recruits soldier-like behavior as a character trait. The facts which military sociologists gain from group behavior and which military psychologists describe as character requirements are the results of education and training in the armed forces. Military pedagogy is thus the logical basis of military psychology and military sociology. The results and facts established by military psychologists and military sociologists should be accompanied by the causal question on the failings of military pedagogy, if studies in the fields of military psychology and military sociology reveal numerous irregularities within everyday military life.

Altogether, it is the task of military pedagogy to reconstruct soldier-adequate behavior. Military pedagogy as a doctrine is orientated towards the concept of peace in freedom in terms of logic, towards international law of war and the UN Charter in terms of ethics and, in terms of praxeology, towards the developmental task for recruits according to the theory of planned behavior. Military pedagogy as historic research describes, analyzes and assesses military curricula and the way of teaching their contents in the armed forces, the core issues being timeless, invariable, and applicable elements of soldier-adequate behavior such as *"loyalty"* and *"honor."*

Military pedagogy as comparative social-science research describes, analyzes, and assesses leadership, training, and education in the armed forces in general, including the impacts of the concept of *"primacy of political authority"* on the armed forces. Basic references for military pedagogy could be: military everyday culture, the system of decisions related to security policy, and the general political culture in terms of military necessities and decisions related to security policy.

Military pedagogy as biographic research deals with the conditions under which the integration of soldiers' personalities is successful, and at the same time with the irregularities preventing such a success.

Application of Leadership Development and Civic Education

In 1993, the German Federal Minister of Defense promulgated the aims and principles of the concept of Leadership Development and Civic Education.[10] The concept is based on the following three pillars: a) The values referred to in the Basic Law of the Federal Republic of Germany; b) the constitutional principles of the Bundeswehr; c) the status of the Bundeswehr integrated into society.

While the Basic Law and the legal bases for the Bundeswehr can be categorized as codified law, the interpretation of which is limited, the status of the Bundeswehr within society is of dynamic quality. The application of LDCE thus

[10] German Bundeswehr Joint Service Regulation 10/1.

gains the flexibility needed to prevent the emergence of rigid behavioral schemes and to avoid that it should become too abstract to be applied in practice.

Keeping the doctrine open for innovative adaptations of military leadership to changes in society is a strength which characterized it from the outset. Initially, *Leadership Development and Civic Education* used to be a topos and was highly suited to reach an understanding between persons who had different, if not divergent, opinions about the Bundeswehr.

The concept of Leadership Development and Civic Education did not become a topos out of the intention to create one. Preuss (1986) gives a likely picture of the situation at the beginning of the rearmament of the Federal Republic of Germany: It was only on the basis of their military education - and none other - that the first generation of Bundeswehr officers thought about what a society worth living in should look like. Their efforts are commendable and deserve our recognition but were bound to fail in terms of sophistication and professionalism, since the officers lacked the necessary prerequisites. The General Committee's negotiation results on *"Inner Cohesion"* during the Himmerode Experts Convention. The meeting was lacking profundity in terms of intellectual structure, for which Preuss gives the following explanation: In a short period of time, former professional soldiers had to discuss and write down topics for which they did not have the necessary competence such as: character formation, education, feelings of the German people, freedom, social justice, ideals, national commitments, European feeling of solidarity, sound love for one's fatherland, European ideals and values, democratic types of state and ways of life, belief in Europe, political and soldierly obligations, ethic education, European history, citizens convinced of their nation, social structure and the people's education, as well as true soldiership.

The *Leadership Development and Civic Education Center*, established as an authority on the subject matter by the Federal Ministry of Defense, is investigating how the concept of Leadership Development and Civic Education, in comparison to international military leadership principles, can be developed beyond the idea of a mere Bundeswehr doctrine. It is pleasing to note that even representatives of foreign armies, who enroll for training courses at the Bundeswehr Leadership Development and Civic Education Center in Koblenz, ask for information about the concept of Leadership Development and Civic Education.

Baudissin, who is sometimes called the founding father of the Leadership Development and Civic Education concept, was right about integrating the army and society via the guiding image of citizen soldiers, and he proved his opponents wrong. Even at the beginning of the emerging new Bundeswehr, while Baudissin's colleagues in the Amt Blank[11] considered the Cold War and the combat strength of the Bundeswehr top priority, Baudissin already emphasized the importance of an intellectual framework. His contributions were essential for the acceptance of the Bundeswehr in Germany and the Western community. Officers of the Bundeswehr recognize the intellectual quality of military leadership particularly when serving in multinational units where they witness how the concept of military leadership is sometimes put into practice in the forces of allied nations.

The goals of Leadership Development and Civic Education in the early days of the Bundeswehr are important in order to establish the Leadership Development and Civic Education doctrine in terms of its field of application and action-regulating effect. An informative interview is cited in Walz (1987) during which Baudissin and Karst talk about Leadership Development and Civic Education. Their answers make clear that Leadership Development and Civic Education was a special and successful solution concerning the problem of rearming postwar Germany, which was considered to be politically necessary. Nevertheless, neither Baudissin nor Karst think about universities and colleges as a medium dealing constructively with that topic and thus securing the progress of their military doctrine. They cannot extrapolate the future of military leadership in a sensible way. In this respect, such a pragmatic idea is lacking the foundation in terms of military pedagogy and educational science. Those who seek short-term cover from the machinations of domestic politics by giving up sensible principles of military order and leveling out the difference between the profession of a soldier and any other job, those who project and advocate this opportunistic misunderstanding of military professionalism have given up their responsibility for the military spirit in favor of the *zeitgeist* which reigned in those days. This includes survey results which turn the meaning of national defense, alliance policy and loyalty into isolated causes. Convictions, moral obligations, subjective norms, and behavioral control, i.e. effective factors of the theory of planned behavior, are generally not taken into consideration.

Without doubt, Leadership Development and Civic Education can be considered a positive concept reflected both in the institution of the Leadership Development and Civic Education Center, and in the form of the relevant Joint Service Regulation. The military pedagogical doctrine designed by Baudissin and his staff received a place in a social organization including the option to promote the armed forces in such a way that they fit nicely into society. This is achieved when all recruits serving in the Bundeswehr and assuming

[11] The 'Amt Blank' (Agency Blank) was named after a senior official tasked with organizing the new German Armed Forces. Later on, this agency became the German Ministry of Defense.

responsibility for public welfare in a democracy experience their duties in a positive way.

Joint Service Regulation 10/1 comprises the fields of application of Leadership Development and Civic Education while also managing to pass information on to the parliamentary public. The area of competence of *"Leadership Development and Civic Education"* breaks up into the following fields: Leadership; Personnel Management; Law and Military Discipline; Welfare and Recreation; Health Service Support; Military Chaplainry; Professional Organization and Training; Political Education; Public Relations and Information Service; and Organization.

LDCE is not structured systematically; rather, each piece is independent from the others, the advantage being that each part can be regarded separately without needing to link the pieces in terms of logic, ethics, and pragmatics. Thus, more than one color is needed to portray the concept of Leadership Development and Civic Education. Since the concept is laid down in a Joint Service Regulation and maintained by the Leadership Development and Civic Education Center, there is scope for numerous training and further education activities. The guiding image of citizen soldiers and the main ideas of the Leadership Development and Civic Education concept can in many cases be referred to serve as a justification or reminder. All sorts of good intentions, in the form of a moral appeal, can still be brought into the discussion about defense policy, the Bundeswehr, and the continued development of the military pedagogical doctrine. (Ebeling; Seiffert; Senger, 2002)

To sum up: The military pedagogical doctrine of *Leadership Development and Civic Education* came into existence because of the negative assessment of the Wehrmacht and its misuse by Nazi Germany. In terms of interior policy, it was intended to reconcile West German postwar society, which rejected all kinds of military service (*Count-me-out-attitude*), with the new German Armed Forces. Civilians and military personnel were supposed to get jointly involved into the concept of citizens in a democratic society. At the same time, this contributed to achieving the goals in terms of exterior policy: The integration of Germany into the Western community and the North Atlantic Treaty Organization.

LDCE and Military Pedagogy

The fields of military pedagogy and Leadership Development and Civic Education should, if properly presented, also be reflected in the comparison of different texts. It needs to be looked at how supporters of Leadership Development and Civic Education define the purpose of this concept and whether there are significant differences to theories of military pedagogy in terms of methods and facts on a higher level of knowledge. The key issues are the diverging purposes: On the one hand, the doctrine of Leadership Development and Civic Education for the Bundeswehr, on the other hand, statements in terms of military pedagogy about armed forces as such.

In a presentation given by Baudissin, the founder of LDCE, on April 28th, 1953 in the city of Siegburg, he claims: (Baudissin 1969: 142)

> *"It is the individual who should be our focus within the European order, it is the individual who should exert all ordering force, it is the individual, however, who also represents chaos and fear and thus anarchy and the temptation to establish a collective way of life. It is within this individual that we find the real theater of war in which the conflict takes place, the individual being at the same time the objective of war and the warrior."*

It is interesting to note that Baudissin's written statements, which he left to posterity, are all similar to this one. Particularly noteworthy is the emphasis he puts on the individual, on humanity, which can also be found in his later statements about citizen soldiers. Baudissin referred to some sort of transformation of the individual towards peace-ableness. The individual exerts the ordering force. The individual, however, represents also chaos, fear, anarchy and the temptation to establish a collective way of life. This sounds like a reflection of existentialist philosophy about the human being thrown into existence. Officers talking about the theater of operations generally refer to the external aspects of the ongoing action, the combat, the opponents, and the battle as an event. Soldiers are expected to be selfless and thus willing to risk their lives according to the reason of the state.

Nobody had that in mind in 1953, given that Hitler's misuse of the principle of order and obedience had led to a general and profound loss of loyalty among soldiers. Both the political opinion and the Bundeswehr still harbor reservations in that respect. It seems, thus, that the purpose of Leadership Development and Civic Education is still to regulate the attitude of German society towards the armed forces by aligning the military to the civilian way of life. Soldiers are citizens as anyone else, they just wear uniforms. No generic reference to military pedagogy can be found in Baudissin's statements.

Let us go forward in time. Nearly 35 years later. The following text has been taken from a presentation given by Ulrich Hundt: (1989: 35-51)

> *"The dissemination of ideals means to meet soldiers where they really are! They live in their own "world of ideals" which is imbued with their attitudes, convictions and values. Something we might consider to be necessary, might not be necessary for them, it might be even completely superfluous. Our ideals might not be necessarily their*

ideals and no one will make them adopt another person's convictions, or even advocate them against their own wills and convictions. At the end of the day, motivation and the adopting of ideals are based on voluntary preparedness."

Reflection about the human being as such is also mirrored in this statement. A commander of the Leadership Development and Civic Education Center psychologizes the motivational nature of the individual who is needed as a soldier but whose ideals and motivations are inaccessible. The author does not state which means and methods would be appropriate to overcome communication barriers. No generic reference to military pedagogy can be found in Hundt's statements.

Contrary to the above mentioned quotes about Leadership Development and Civic Education, which give a mere description of a soldier's inner life, Clausewitz preferred a way of thinking in his works going beyond that. He avoided the description of the individual's state of mind in military scientific texts, although he was quite able to explain them in writing, as Schössler (1991) demonstrates in his monograph on Clausewitz. The fact that the individual characteristics of soldiers are left out of consideration promotes the process to abstract which enables people to generalize on a higher level of abstraction and to make military plans as well as to give military orders. Even today, Clausewitz makes still an impression due to his ability to establish theories, his knowledge of military science and also due to his incorruptibility. In the fifth chapter of his work *"On War,"* under the title of Military Virtues of the Army, Clausewitz states: (Clausewitz, 1831: 364)

"There are only two sources for this spirit, and they must interact in order to create it. The first is a series of victorious wars; the second, frequent exertions of the army to the utmost limits of its strength."

This historical situation has to be put into the present context. While victorious wars were comprehensible goals of those times, things have changed nowadays. Today, success in a military conflict is seen more and more as the safeguarding of peace by deterring military or terrorist threats. UN missions, if successful, have a positive impact on military personnel's readiness for action.

Clausewitz's statements are not only legitimated by their historical substance but they are also important with regard to military pedagogy. The success experienced in a military operation depends on whether the armed forces were able to meet the demands made on them. Both factors constitute the moral spirit of the army. Today, we may assume this to be the combat effectiveness of a military unit experienced by the soldiers, i.e. their collective experiences of their unit's effectiveness. The combination of three elements: positive military experiences, training for readiness, and the personnel's fighting morale fortified by successful experiences, produce a unified whole: the soldier's image. The reconstruction of Clausewitz's statements can be regarded as a model in terms of military pedagogical philosophy.

Baudissin focuses on the problems of the human being's individual character, i.e. he pursues the philosophical path whereas Hundt points to the communication problem of a human being whose ideals are not easy to bring in line with military ideals. A program of measures adopted by the Leadership Development and Civic Education Center pursues the same path. Clausewitz tackles this issue related to military pedagogy in a different way. He prefers the way of thinking in form of a model, as a consequence of which his statements are much more suited for generalization than are statements about the military and society referring to the here and now. In the fourth chapter of his work *"On War"* titled *"The Principal Moral Elements"* he makes a statement about three variables of the model: the talents of the *feldherr,*[12] the combat experience of the troops, and the patriotic spirit. If the *feldherr's* possibilities to take action are determined to a large extent, the following conclusion arises:

> *"It cannot be denied, therefore, that as things stand at present proportionately greater scope is given to the troops' patriotic spirit and combat experience."* (Clausewitz, 1831: 359)

The three variables are all interacting. Clausewitz also assumes this kind of interaction when saying that the Feldherr gains possibilities to take action if supported by a patriotic spirit and an army experienced in combat. The interactive relationship is also visible in the following quote:

> *"A people and nation can hope for a strong position in the world only if national character and familiarity with war fortify each other by continual interaction."* (Clausewitz, 1831: 370)

If one links these three variables of the model the result will be the following triangle:

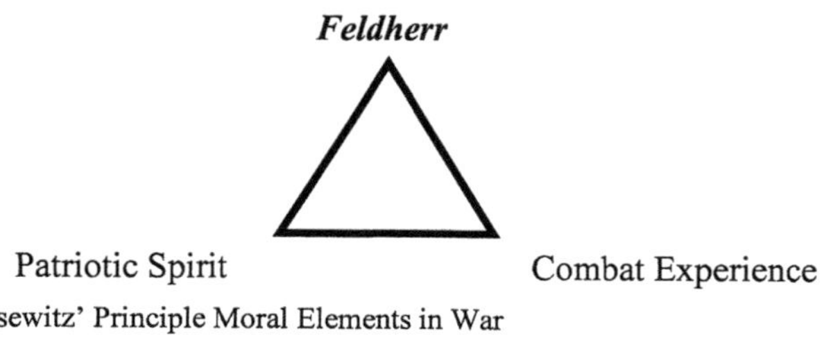

Fig. 2: Clausewitz' Principle Moral Elements in War

[12] The German term "*Feldherr*" is a historical expression for "commander in chief."

The relationship between the national character of a people, a nation, and war could be translated into and presented in the graphic below:

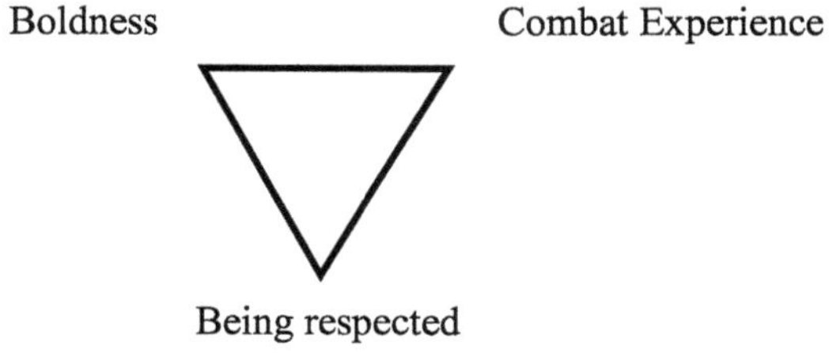

Fig. 3: National Character and Familiarity with War

"A strong position in the world" is translated with *"being respected."* Without a doubt, reference is made to the universalistic approach in terms of security policy in general and to the necessary mental prerequisites in particular. The concept of *"national character"* is interesting from a military pedagogical point of view, since it makes allusion to the idea that all citizens of one nation could agree on an intellectual basis upon basic principles in terms of security policy and to the prerequisites needed for maintaining peace in freedom. A bold nation is daring, as is a bold Feldherr.

Today, we would rather speak of *"being prepared to take risks"* than of boldness. Neither a Minister of Defense nor a commander of a military district nor a commander of a UN unit would respond directly if addressed by *"Feldherr."* All of them know the function a high ground for Feldherren had, and the monocular telescopes with which the Feldherren used to observe the course of the battle from a distance are exhibited in army museums.

Nowadays, the course of the operation can be displayed on screen thanks to the digitalized command systems of the armed forces, which are supported by satellite communications (SATCOM) and the Battlefield Management System (BMS). The overall responsibility for the military course of action still exists today. Commanders' training in information management, the way they learn how to take calculated risks, thus is a top priority in military pedagogy.

The expression *"combat experience"* or, as Clausewitz literally put it, *"familiarity with war"* might be irritating from today's perspective. Neither Clausewitz in the past nor today's sensible commanders are interested in shedding blood. Peace, however, is still in peril and wars have not been eradicated yet. The armed forces of nations living in peace cannot dispense with training under combat conditions. During UN peace-keeping missions it is also allowed to use weapons for self-defense or to protect others. The official

Bundeswehr personnel assessment form now also requires rating officers to comment on the rated person's *"performance in field missions."*

Although military missions are carefully planned and carried out the possibility cannot be ruled out that soldiers may die or be wounded during a mission. Some supporters of the Leadership Development and Civic Education concept will freely admit to that. The reference to other nations where the armed forces do have a more or less unbroken military pedagogical tradition is a good justification with regard to the pending Bundeswehr reorganization and the restructuring of leadership, training, and education for out-of-area missions. The thesis is that military pedagogical ideas are suited to supplement the military pedagogical doctrine of Leadership Development and Civic Education by giving a generic picture about military personnel's preparedness for action and acceptance to take risks in the form of a theory of soldiership and its application in everyday military life.

Military Pedagogy as Scientific Systematology

Military pedagogy as a partial discipline of educational science is an institutional pedagogy for the armed forces, just like school pedagogy is an institutional pedagogy for schools. The concept of Leadership Development and Civic Education is a special doctrine in the field of military pedagogy, i.e. for the Bundeswehr. Since the concept is a doctrine, a theory of military pedagogy is required in order to catch up with military pedagogical doctrines of the armed forces of other nations. In this respect, the question could also be how the armed forces of other nations could benefit from the Leadership Development and Civic Education concept. Military pedagogy is constituted by the fact that the armed forces are the core on which educational methods, i.e. hermeneutics and empirical science, are applied. All doctrines need reference to a scientific subject, since they would otherwise run the risk at losing touch with their goals through senseless action routines, institutional disruptions, and irritating political exertion of influences. This point is especially stressed by Toiskallio (2002: 87) who considers the compensation of the technological tendency of 'cyborgization' as the important function of Military Pedagogy. Military training and education always has to allow for the difference between man and machine, even if automatized behavior and the use of highly sophisticated weapons may be necessary for survival.

Military pedagogical ideas, referring to the computer-aided training of noncommissioned officers, are necessary if course participants start to search the training material for videos only and to replay the soundtrack by pressing the rewind button over and over again. If the instructors threw a glance at the evaluation protocols, which indicate that the participants did not reach the learning goals, they would notice that they cannot leave the participants by themselves in the computer room. The instructors bear the responsibility for a successful training during the whole course.

Soldiers learn techniques of behavioral training. Examples reflecting real life are taken from military day-to-day life. However, if you have to neutralize the recruit's fear of rappelling down a steep precipice, you also have to convince him of the necessity to risk life and limb. Group-dynamic techniques and leadership counseling often ignore the importance that norms, convictions, and moral obligations have with respect to the training and education of soldiers, regardless of all behavioral modifications. While ideas of behaviorism might be helpful in military training, they cannot in any way substitute the cultural substance of soldiership relating to military pedagogical basic concepts such as constitution, democracy, and freedom (Fig. 4a). This has to be seen in connection with the geopolitical development of security policy, by which national armies in transition become embedded in a system of global alliances. In this way, missions of armed forces are ever more dependent on UN mandating.

Establishing Military Pedagogy at Universities

The concept of Leadership Development and Civic Education as a doctrine of military pedagogy, grouped within the field of military pedagogy as its higher scientific discipline, claims to constitute military pedagogy as such. This is the goal of the working group Military Pedagogy, which is affiliated with the NATO group on training and education. So far, university teachers giving lectures at public universities could deal with matters of the armed forces only as a secondary occupation. Military pedagogy could not be introduced yet as a subject at Bundeswehr universities.

Although social change is something that is widely discussed by many people, the attitudes of educational scientists towards the armed forces and soldiership within democracy have remained extraordinarily distanced. The various Federal Ministers of Defense, understandably seeking to keep the political waters calm, could not yet afford to ask for the support of educational scientists concerning an institutional pedagogy for the armed forces. Manfred Wörner, former Federal Minister of Defense, tried to gain the support of the Ministers of Education and the Arts of the *Länder*[13] for a common statement about how to deal with questions of security policy in schools. His attempt ended in a spectacular failure. (Lutz, 1984) Against this backdrop, the military pedagogical initiative which was initiated by vanTrotsenburg (1989) at Klagenfurt University, supported by the Austrian Federal Ministry of National Defense (Jung &

[13] In Germany as a Federal Republic, the *Länder* can take independent decisions in matters of education and culture.

Florian, 1994) paved the way for the conferences in Helsinki (2000), Stockholm (2001), and in Zurich (2002) is even more welcome.

Even if it has not been possible to date to establish a chair of military pedagogy - particularly one that focuses on Leadership Development and Civic Education - at the initiative of a university, it might be appropriate to approach a foundation for financial support and to look for a university that would establish such a funded chair. This would enable the development of a partial discipline of educational science which deals, in the fields of research and teaching, with a topic all citizens would benefit from, since the armed forces contribute to the fact that freedom and the rule of law do not stand helpless in the face of an external threat by protecting the territory of the Federal Republic of Germany, by being a member of NATO, and by advocating the goals of the United Nations. The Bundeswehr protects the constitution, which guarantees basic rights, including the right of conscientious objection, freedom of speech even if it is used to the disadvantage of the armed forces, and peace-striving pedagogical activities of all kinds.

Conclusion

The concept of Leadership Development and Civic Education is a topic of military pedagogy, the latter being a partial discipline of educational science. This is the conclusion which can be drawn from the comparison of both fields of applications of the subjects to be combined. In this respect, the armed forces of civilized nations have a right to ask for their own institutional pedagogy given their global responsibility for peace and freedom. This institutional pedagogy is also legitimated by the fact that the development and the maintenance of military everyday life depend on research in the field of military pedagogy. Furthermore, it is the task of military pedagogy to reconstruct soldiership within democracy whenever deficits in leadership, training, and education of the armed forces come to light, including deficits noticed in the organizational development of the armed forces. It should be examined whether the means and methods of special military pedagogical doctrines such as the Bundeswehr Leadership Development and Civic Education concept can be generalized. We should answer whether the goal of the Leadership Development and Civic Education concept can be transferred to the armed forces of other nations or whether it could be reshaped in order to be applicable in other armed forces. Given the military cooperation in multinational units, the need for such answers is already visible.

It would also be appropriate to establish military pedagogy as a scientific discipline at universities since this is the place where teaching and research can be carried out in freedom. The only limitations arise from the ethical responsibility for scientific behavior; there is no room for the executive authority of superior officers or party political influences on the construction of theories.

Concerning the development of this under-represented scientific subject, it is fortunate that, despite all resistance, a chair of military science and security policy has been established at the Bundeswehr University in Munich after 20 years of struggling. We still need more. The military would benefit from that, as it would contribute to the public granting them the professional respect they deserve. The integration of the armed forces into society in the Federal Republic of Germany will be most successful if we reduce prejudices against military pedagogy. Sensible attitudes in terms of acceptance of security policy and the armed forces are predominantly promoted and spread where freedom of teaching and research is the rule.

The orientation of military pedagogy towards an international approach smooths the way for developing standards in leadership, training, and education which have the common goal to maintain peace in freedom. Military pedagogy, from a universalistic point of view, can provide access to information - in the sense of international military ethics and through teaching and research - about the soldierly image of the individual for all armed forces and about the common striving for peace around the world.

While such effectiveness of military pedagogy is still to be awaited, today's situation in the world makes the following a matter of course: If you want peace, be prepared to defend it.

References

Ajzen, I. & Madden, J.T. (1986): Prediction of Goal-directed Behavior: Attitudes, Intentions, and Perceived Behavioral Control. In: J. Exp. Soc. Psychology, 22, 453-474.

Baudissin v., W. (1969): Soldat für den Frieden (Soldier for Peace). München; Piper.

Bundesministerium der Verteidigung (1993): Zentrale Dienstvorschrift 10/1. Bonn, Federal Ministry of Defense: Joint Service Regulation 10/1.

Clausewitz, Carl von (1831): Vom Kriege (On War). W. Hahlweg (ed.) (1991): Bonn, Dümmler.

Ebeling, Klaus; Seiffert, Anja & Senger, Rainer (2002): Ethische Fundamente der Inneren Führung (Ethic Foundations of Leadership Development and Civic Education, Bundeswehr Institute for Social Sciences). SOWI-Arbeitspapier 132, Strausberg, Sozialwissenschaftliches Institut der Bundeswehr.

Hundt, Ulrich A. (1989): Die gesellschaftlichen Bedingungen des Wehrdienstes und unsere Folgerungen (The Social Conditions for Military Service and Our Conclusions). In: Kister, K.; Klein, P.: loc. cit. 35-51.

Huntington, Samuel P. (1957): The Soldier and the State. Cambridge, Mass., Belknap Press.

Janowitz, Morris (1971): The Professional Soldier. New York, Free Press.

Jung, Hermann & Florian, Heinz (1994): Grundlagen der Militärpädagogik (The Basics of Military Pedagogy). Frankfurt/Main, Peter Lang.

Kister, Kurt & Klein, Paul (eds.) (1989): Staatsbürger in Uniform - Wunschbild oder gelebte Realität? (The Citizen Soldier - Wishful Thinking or Reality?) Baden-Baden, Nomos.

Lutz, Dieter, S. (ed.) (1984): Weder Wehrkunde noch Friedenserziehung? (Neither Military Science nor Peace Education?) Baden-Baden, Nomos.

Moskos, Charles (1977): From Institution to Occupation - Trends in Military Organization. In: Armed Forces and Society, Vol. 4, 41-50.

Opitz, Eckardt (ed.) (2001): 50 Jahre Innere Führung. 50 Years of Leadership Development and Civic Education. Bremen, Edition Temmen.

Paschen, Harm (1979): Logik der Erziehungswissenschaft (Logics of Educational Science). Düsseldorf, Schwann.

Pöggeler, Franz (1989): Probleme der Ausbildung in demokratischen Armeen (Problems of Training and Education in Democratic Armed Forces). In: v. Trotsenburg, E. (ed.): loc. cit., 87-113.

Preuschoff, Klaus-Jürgen (1988): Suizidales Verhalten in deutschen Streitkräften (Suicidal Behavior in German Armed Forces). Regensburg, S. Roderer.

Preuss, Helmut (1986): Bundeswehr und Innere Führung (Bundeswehr and Leadership Development and Civic Education). Siegburg, Selbstverlag.

Rosen v., Claus (1981): Bildungsreform und Innere Führung (Educational Reform and Leadership Development and Civic Education). Weinheim, Beltz.

Royl, Wolfgang (1992): Euro-militärische Identität als Entwicklungsaufgabe (Euro-Military Identity as Developmental Task). In: Dexheimer, W.; Royl, W. (eds.): Die pädagogische Mitverantwortung für die Sicherheitspolitik der freien Welt (The Pedagogical Co-Responsibility for Security Policy of the Free World). Baden-Baden, Nomos, 199-239.

Royl, Wolfgang (2002): Euro-Military Identity in Transition. In: Toiskallio, Jarmo et al.: Cultures, Values and Future soldiers. Series 2, No 11, Helsinki, Finnish National Defence College, 13-45.

Royl, Wolfgang (2002): Military Identity under Pedagogical Implications. In: Florian, H. (ed.): Military Pedagogy - an International Survey. Frankfurt/Main, Peter Lang, 189-209.

Schmähling, Elmar (1991): Ohne Glanz und Gloria (Without Grand Style). Düsseldorf, Econ.

Schössler, Dietmar (1991): Carl von Clausewitz. Hamburg, Rowohlt.

Toiskallio, Jarmo (2002): Cyborgs and Humans. Two Paradigms of Military Pedagogy. In: Florian, Heinz: Military Pedagogy - an International Survey. Frankfurt/Main, Peter Lang, 83-100.

Trotsenburg, Edmund van (1989): Militärpädagogik (Military Pedagogy). Frankfurt/ Main, Peter Lang.

Walz, Dieter (ed.) (1987): Drei Jahrzehnte Innere Führung (Three Decades of Leadership Development and Civic Education). Baden-Baden, Nomos.

Outi KALLIOINEN, Häme Polytechnic Delevopment Uniz, Hämeelinna

ETHICAL DIMENSION IN TEACHING AND LEARNING

Introduction

The purpose of this article is to reflect on the ethical dimension in teaching and learning within military training. Cultural, social, and historical issues as well as multicultural aspects are reality in today's Europe, and military teachers' and individual soldiers' ethics and morality are challenged in various situations every day. It is therefore essential to create a solid ethical foundation for military teachers' professional growth in the army.

Toiskallio (2000: 6-9) emphasizes that soldiers and officers are cultural beings and actors in different and changing environments such as organizations, peacetime communities, battlefields of war, in peacekeeping and crisis management operations, as well as in societies and nations. This setting creates a great challenge for all military training and education institutions.

Teacher's Professional Growth

The challenge of a teacher's professional growth emphasizes the importance of each teacher's personal expertise. The expert teacher participates personally in life-long learning, understands the change in teaching and learning, utilizes the possibilities for learning at workplaces and the possibility to use information technology in learning. He is also capable of creating innovative learning environments (e.g. simulations and projects) in a productive cooperation with colleagues and local networks in different working places. The ideas of theory and practice in learning are dynamic and they have a great effect on the teacher's role as well.

Helakorpi has especially outlined the expertise in vocational education in his article *"The Trainer's Expertise."* (Helakorpi, 1999: 378-388) According to him, the vocational trainer can be understood as an expert in training whose know-how is constantly developing and permanently influenced by the change in working life as well as by the change in concepts of knowledge and learning. The vocational trainer is also instructing the trainee in growth and professional socialization, which highlights expertise and the trainer's responsibility of the trainee's individual growth and ethical development. Socialization has an important function in every training institution, including military training, and therefore it is important that the trainers feel responsible for the trainees' individual growth and ethical development.

A teacher's professional development is generally also supported by the teaching community. Bergen and Engelen (2003: 95) stimulated teachers' professional development in their research by peer coaching in secondary education. They audio-taped 45 coaching dialogues and analyzed the data qualitatively using *KWALITAN*, the computer support program for qualitative analysis. In their paper they argue that coaching is a powerful learning opportunity for both parties.

Coaching as a concept is defined by Bergen and Engelen in the following way:

> *"... a form of professional collaboration and support to improve professional development and craftsmanship through experimentation, reflection, exchanging of professional ideas, and problem solving. Peer coaching is the process by which, on the basis of mutual trust, two or more colleagues co-operate in order to reflect on their own practice, to try to acquire, extend and refine (new) skills, to exchange ideas, to teach each other, to conduct action research in their classrooms and try to solve problems encountered in the workplace."*

According to Showers and Joyce (1996: 12-16) there are three objectives for peer coaching: (I) Teachers cooperate for improving their own professionalism; (II) Teachers develop a common language and framework of ideas that are necessary for an analysis of the new knowledge and skills; (III) Mutual coaching provides a follow-up structure, which is essential in acquiring new skills and strategies.

The work on teachers' own professional development, staff development, and school development is also stimulated by peer coaching. Peer coaching can be seen as a powerful intervention in stimulating teachers to reflect on their own professional performance practices in order to find out what constitutes good teaching. Ethical issues are common in teaching and training and there should be a constant dialogue on ethical matters and questions within the teaching community. The idea of peer coaching is constructive and could easily be organized in a systematic way as part of the progress of teachers' professional development in military training institutions. Teachers and trainers are used to working together as a team nowadays. As the doors of the classrooms are opening up, a good starting point for organizing peers is coaching. Modern learning environments also support peer coaching. Peer coaching could provide a natural forum for a dialogue on ethical issues, given that there is enthusiasm and good will.

Social Interaction in Producing Knowledge

Teaching military ethics can be seen as a professionally all-inclusive and challenging communication process. Teachers and students are heading for the learning targets and creating a confidential learning atmosphere, in which not only the teacher's own personal growth and pedagogical expertise, but also interrelationships, trustworthiness, and knowledge base are emphasized.

From a constructivist point of view, the teacher is not only demanded to understand the issue he is teaching and to possess the required skills, he should also be able to understand and support the individual learning processes that play a role at various stages and are based on various starting points. In addition to mastering the substance he should manage the theory of the teaching-learning-process so that he can adapt and specify the curriculum by reflecting the learning situation and drawing everybody's attention towards the learning goal. In managing teaching as a communication process, it is required to understand social interaction and to have the necessary skills for it. (Rauste-von Wright and von Wright, 1994)

In the constructivist view learning attention is drawn to the student's active self, so that knowledge is not simply transferred to the recipient but rather every student creates and produces knowledge in social interaction of the learning community. In the discussion culture of learning situations the student can analyze and reflect his/her own thoughts and better understand the whole picture by reflecting other students' opinions. It is important that every student is given the opportunity to participate in the conversation and be heard. The student's learning process is remarkably enhanced through social interaction. Other learners' experiences may also enlarge and challenge the concept of knowledge and improve an individual student's problem solving skills. (von Glasersfeld, 1995)

The teacher's role is crucial in facilitating the interaction and directing the conversation towards essential matters in military ethics. Raising fruitful and problematic questions is often much more important than giving specific answers. On these occasions the teacher's professional and pedagogical expertise is challenged. The function of feedback in learning is not to point out or highlight mistakes, but to enhance learning and use the mistakes as a means for learning and problem solving. However, it is important to remember that this does not mean missing the purpose of learning. The students easily notice if the teacher's own ethical foundation is weak, substance knowledge is not current, or if the psychological learning climate is ruined by lack of authority or the teacher's negative attitudes.

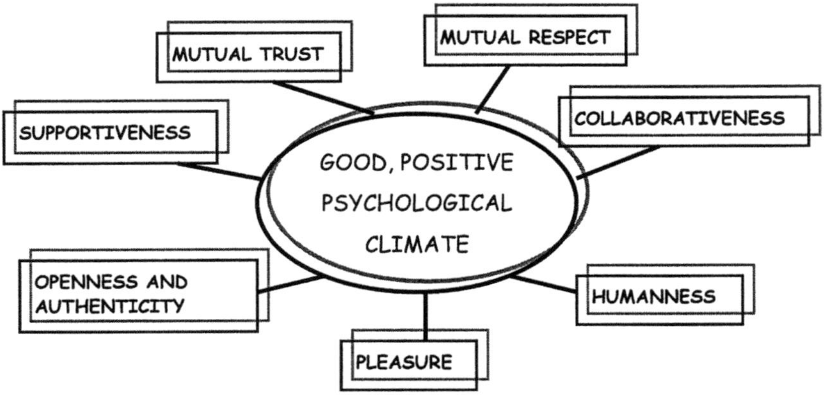

Figure 1: Positive psychological climate (adapted by Knowles et al. 1984, 15-17).

A positive psychological learning climate (Fig. 1) is an essential element for social interaction in learning. If the students do not feel safe and respected by the teacher in the learning situation, their learning is prevented and their energy for learning flows into a wrong direction - towards their own negative feelings. A negative learning climate greatly prohibits and suffocates social interaction. Every single student is worth attention and in the group it is obvious if someone is given no attention. Taking care of the students, tutoring, coaching, and instructing each student individually if necessary, creates trust and respect which supports the teacher's real authority. An expert teacher is a creator of a good learning atmosphere and he treats students equally and with respect, which is a sign of high performance in professional behavior.

Meeting the learners' needs individually will create learning environments that would enhance good and efficient learning. This also means that the teacher is able to recognize different kinds of learners and takes them into consideration in learning situations. The learners are responsible for their learning but the teacher is responsible for providing the proper atmosphere where the students are willing to take responsibility for learning and where they want to learn the things that are set as targets. The teachers are educators, but while educating others one grows as well. In this sense learning situations can be considered as forums for mutual growth.

Ethical Dimension in Teaching

Every teacher is somehow involved in the process of life-long learning and part of it is the constant reflection on the ethical dimension in teaching, learning, and educational guidance. During the teaching career every teacher faces situations and problems in which the individual teacher's concepts of human beings, values, and ethical codes are challenged. There is no handbook for resolving difficult problems and often challenging and complex situations. The handbook for these situations is produced during the personal and professional growth of the teacher in teaching communities. Therefore the atmosphere for open dialogical interrelationships is needed within these teaching communities. As teachers we all make mistakes and these mistakes should be seen as possibilities for individual and mutual learning and development.

The most important message of ethics cannot be revealed only by performing culture studies or through reflecting the values that support practical purposes. Being ethical has connections to such concepts as metaphysics and ontology - our views of reality, being, and knowledge. Reflecting what is ethical and what is good has a self-valued position because it is a discussion of the nature of the good. The mission for science and philosophy is to examine the construction of good and right; by doing so the level of ethical conscientiousness can be raised. (Juuti, 2002)

When cultures meet and rapidly mix through internationalization it is not enough to have narrow technical-rational standards or to strictly stick to the 'right' views of one's own culture. As values and conceptions of good are often culture-related, cultural sensitivity should be increased. The mission for science and philosophy is the ability to open the cultural code and the specific habits, in which the good is constructed. (Juuti, 2002: 10)

In teacher education and in teaching communities it is essential that teacher students and teachers think deeply about their value basis and their ontological and epistemological standpoints together in a dialogue with experienced teachers. On the basic level the concepts of the human being and his next surrounding world are directing the teacher's professional behavior wherefore it is valuable to reflect up on these concepts in order to recognize the roots of individual values. The foundations of a teacher's action are attached to the value basis and a teacher's ethical code is like a reflection of the value basis. In each teacher's moral actions his ethical code becomes visible.

From the humanistic point of view, one of the core values in a teacher's work and in school leadership is caring and respect for others. Caring requires

> *"fidelity to persons, a willingness to acknowledge their right to be who they are, openness to encountering them in their authentic*

individuality, a loyalty to the relationship [...] This value is grounded in the belief that the integrity of human relationships should be held sacred." (Leithwood, 1999: 212)

Freedom, equality, and social justice can also be considered as core values and they are rooted in Dewey's concept of a democratic school.

Jensen (2003) has studied the exploration of critical self-reflection in the teaching of ethics for physical therapy students. There is obviously a rising demand for teaching ethics, especially in the area of humanities and in health professions, but the course contents of professional ethics should be integrated in a more meaningful way into the curricula in general.

In this complex world the rising demand for teaching ethics is a general challenge throughout the educational system and in military education as well. This idea of integrating ethics applies also to teaching military ethics. Integrating military ethics into the curricula in general provides a fruitful standpoint for the students' professional and personal growth. Integration brings practice and theory together and thus the learning becomes more meaningful for the students.

Purtilo has used an appropriate metaphor when saying teachers are doing *"ethical planting for future generations."* (Purtilo, 2000: 1112-19) It seems as if this metaphor can be widely adapted to other professional fields as well and not only in health care education. Jensen asks the teachers about how they facilitate student awareness and insight into social, cultural, ethical, and moral aspects of practice and this question can be seen as crucial in the development of the entire society.

Dewey argued about the object and reward of learning which is continued capacity for growth where students *"develop skills and habits of mind that will enhance their creativity and problem solving abilities with respect to the issues they are likely to meet."* Authors Jensen and Paschal call for tools of ethics that provide reflection on complex situations in daily practice. (Jensen & Paschal, 2001: 42-47) For teachers in a professional education environment it is a necessary but difficult challenge to facilitate the reflective habits of the mind. However, to be able to facilitate reflection on ethical issues, the teacher has to constantly reflect on one's own perspectives and thoughts in a life-long process of professional development and growth.

Professional military teachers should possess high ethical standards which can be seen in the teachers' attitudes and actions. A good teacher is an exemplary role model for the students and plays an important role in teaching military ethics.

References

Bergen, T. & Engelen, A. (2003): Looking for the Quality of Coaching Dialogues. Paper presented at the annual meeting of the American Education Research Association, Chicago.

Helakorpi, S. (1999): Kouluttajan asiantuntijuus. Kasvatus 4/1999, 378-388.

Jensen, G. M. (2003): Exploration of Critical Self-Reflection in the Teaching of Ethics. The Case of Physical Therapy. Paper presented at Division I, Education in the Professions, at the annual meeting of the American Education Research Association, Chicago.

Jensen, G.M. & Paschal, K. (2001): Habits of the mind: student transition toward virtuous practice. Journal of Physical Therapy Education, 14 (3), 42-47.

Juuti, P. (ed.) (2002): Ethosta etsimässä. Puheenvuoroja johtamisen ja yrittämisen etiikasta. Aavaranta-sarja n:o 50. Juva, Bookwell.

Knowles, M. et at. (1984): Andragogy in Action. San Francisco, Jossey-Bass.

Leithwood, K.; Jantzi, D. & Steinbach, R. (1999): Changing leadership for changing times. Buckingham, Open University Press.

Purtilo, R. (2000): Thirty-first Mary McMillian Lecture: A time to harvest, a time to sow: ethics for a shifting landscape. Phys The, 80, 1112-1119.

Rauste - von Wright, M. & von Wright, J. (1994): Oppiminen ja koulutus. Juva, WSOY.

Showers, B. & Joyce, B. (1996): The Evolution of Peer Coaching. Educational Leadership 54(3), 12-16.

Toiskallio, J. (2000): Introduction. Is There Military Pedagogy? In: Toiskallio, J. (ed.) Mapping Military Pedagogy. Finnish National Defence College, Department of Education, Series 2, No 7, 6-9.

Glasersfeld, E. von (1995): A Constructivist Approach to Teaching. In: Steffe, L.P. & Gale, J. (eds.): Constructivism in Education. New Jersey, Hillsdale.

Jarmo TOISKALLIO, National Defence College, Helsinki

MILITARY ETHICS AND THE TECHNO-CULTURE OF THE INFORMATION AGE

> *"The security environment continues to change rapidly, bringing new and more diverse risks, challenges and opportunities. (There is)... the enduring requirement for high intensity warfighting skills in order to deter and, if required, fight and win battles. ...*
>
> *But we must also be ready for a wide spectrum of operations, including the growing number of coalition peace support and humanitarian operations. ... we will need to manage carefully the balance between training for warfighting and non-warfighting. Individual training must provide the basic skills that can support warfighting and high intensity operations, but needs to be supplemented to meet other scenarios. The Kosovo campaign also reinforced the need for a more integrated approach to crisis management, supported by common training for personnel from MOD, other Government departments and non-governmental bodies."*
>
> (Modernising Defence Training: Report of the Defence Training Review. The British Ministry of Defence)

The quotation above represents the new way of thinking that is typical today for all Western democracies: *"... we will need to manage carefully the balance between training for warfighting and non-warfighting."* (Donnelly, 2000; Evans & Ryan, 2000; Hall & Fox, 2002) The challenges of security have changed and the military is forced to look for a new culture. Military identity is confronted by a new situation. The problems of the new situation have much to do with the Information Age and its *"technological forms of life."* (Lash, 2002) In order to understand these new *"forms of life,"* we need new approaches like *"cyberpsychology."* (Gordo-López & Parker, 1999)

Together with identity and culture, there is a need to think thoroughly about ethics: not only about the military ethics, but the ethics of security and defense altogether. The American Major General (ret) Buckingham (2002) says it clearly: *"Whatever protects and enhances human life is good, and whatever destroys or degrades human life is evil."* This means that in its purest and most fundamental essence, the purpose of military force is to protect life, not to destroy life. (Glover, 2001)

The Russian Iouri Noskov also underscores a sound ethical leadership in the military. Noskov defines it as follows: *"The professional and moral duty of military officers is to arm their junior officers and younger soldiers with basic moral principles they can rely on to make the right moral choice in complicated situations. This is the meaning of ethical leadership."* (Noskov, 2002)

Perhaps it is just the ethical leadership that is the key factor in a new military culture. Ethical leadership, of course, cannot be managed by computers and artificial intelligence, the highly valued elements of information society.

There are some alarming features in the technologically developed western world; one of them seems to be a growing carelessness among people as a kind of postmodern city/techno-culture. It is connected with severe identity problems with difficulties to find any meaning for life. Noskov sees this clearly when he describes the extremely difficult circumstances in the Russian society and defense forces:

> *"In the sphere of morality, moral degradation resulted in devaluation of such basic ethical categories as 'conscience,' 'duty,' 'honor,' and 'dignity' ... we can say that at the present time, moral security is under a threat in Russia ... moral crisis in modern Russia is accompanied by the growth of crime, prostitution, drug addiction, and the number of HIV-infected people."* (Noskov, 2002)

But we should not be blind to the possible moral crisis in the Western world. In other words, ethics should be reflected to a larger extent if we want to understand the developing new military culture.

It is the task of military pedagogy to conduct research on military training and education. But it would be a fatal mistake to restrict the interest only to training methods. Military training and education always take place in the contexts of societies and cultures. The identities and meaning perspectives of people - soldiers as well as civilians - form the strategic basis for defense. It is just this strategic basis, or center of gravity, that is the most important area of inquiry for military pedagogy. The task is to inquire into the question how do we develop people's growth (*"growth"* means here quite the same as the German term *Bildung*; it refers to identity, morals, and culture).

The concepts of identity, culture, and ethics belong together. Culture can be seen as an order- and meaning-creating activity of human beings. In a culture it is a question of the meaning perspectives of people. When discussing meaning perspectives, we are talking of the ability of people to experience the world and their engagement with it as meaningful. Identity has a close relation to culture: identities refer to the processes of construction of meaning on the basis of cultural attributes; identities are sources of meaning for the actors themselves. (Castells, 1997: 6-7)

According to Manuel Castells, identities organize the meaning, and he defines meaning as *"the symbolic identification by a social actor of the purpose of her/his action."* In other words, our identity has a central role in our action. The young people and little children of today, the potential soldiers in the future, are constructing their identities within the cultures of today and in the value environments we are constructing just now. That is why we should be interested in the way we are making the future culture the strategic center of gravity for the future defense at the national, at the European and regional, and at the global levels. Education is not only teaching some things and skills. Education makes the future culture. It is just in this meaning that we should understand military pedagogy as an area of research and theory on education.

Like Zygmunt Bauman (2001) says, culture is more an active process of making than a passive habitual structure. What kind of directions and purposes do we want for this process of making? What kind of values do we prefer? For instance, Vesa Karonen (2002), a Finnish journalist who specializes in war literature, is critical with the new, emerging literature, which describes superficially and with an admiring mind the activities of mercenaries - hired soldiers - in the battles of the Balkans. Karonen is worried about the quality level of war literature, which should also be seen as a problem for and danger to military culture.

Adopting an active view on culture, a well argued normative basis is needed in order to discuss the future military culture. Wolfgang Royl gives an important basis from a European point of view:

> *"The development of the identity with the military requires that the soldiers may be convinced that the society in which he lives and the global perspective it is gaining have a positive attitude towards peace in freedom... "* (Royl, 2002)

One of the major problems of today is that cultural changes are inexorably linked with technology. Technology is not at all a 'bad thing,' but as Marshall McLuhan said in the 1960s, electronic technology as a medium transforms the nature of social relations - all media transforms us completely.

Almost forty years later United States Army Colonel K. Nygren, when analyzing army developments, says that *"technological innovation is the primary cause of a growing stress in society ... Technological innovation continually remakes the world and is central to the economic process and cultural beliefs - commonly labeled progress - that drive modern civilization."* (Nygren, 2002)

When we are trying to think of the cultural context of the future soldiers, it is just the technological implications - especially those of the information and communication technology (ICT) - that should be in the core area of interest.

Potential future soldiers are the children and teenagers of today. Their senior officers are the college and university students of our present time. Tomasz Goban-Klas calls all of them *"SMS-gener@tion,"* because the use of SMS (short messages) by mobile phones is so dramatically changing not only their styles of communication, but also the psychological characteristic of the whole generation:

> *"... the new SMS-gener@tion has a shorter attention span than the older, [...] able to read sophisticated messages, but it lacks patience and ability to follow logical arguments. It is [...] well adapted to bits of information, and much less to knowledge collection and creation. The third element in the triad - wisdom - is evidently beyond pursuit or desired. Instead, entertainment and togetherness are constantly sought."* (Goban-Klas, 2002: 226)

This young generation described by the researcher of communication and media seems to have a good potential for the *"digital decision making"* of techno-armies, and some of them are perhaps the future *"digital generals."* (Harig, 1996; Dunlap, 1999; Adams, 2001)

While this generation is adept at applying technological tactics, Goban-Klas writes that they lack wisdom: this might mean that these *"digital generals"* might be ethically unfit for command.

References

Adams, T.K. (2001): Future Warfare and the Decline of Human Decision-making. Parameters, Winter 2001-2002.

Bauman, Z.; Tester, K. (2001): Conversations with Zygmunt Bauman. Cambridge, Polity Press.

Castells, M. (1997): The Information Age. Economy, Society and Culture, Volume II. The Power of Identity. Oxford, Blackwell Publishers.

Donnelly, C. (2000): Shaping soldiers for the 21st century. NATO-Review, Summer/Autumn.

Dunlap, C.J. (1999): Technology: Recomplicating Moral Life for the Nation's Defenders. Parameters, Autumn.

Glover, J. (2001): A Moral History of the Twentieth Century Humanity. London, Pimlico.

Goban-Klas, T. (2002): SMS **Gener@tion** or the Emergence of Mobile Media People. In: Pantzar, E. (ed.): Perspectives on the age of the information society. Tampere, Tampere University Press.

Hall, R.; Fox, C. (2002): Rethinking security. NATO-Review, Vol. 49, No. 4.

Harig, P. (1996): The Digital General: Reflections on Leadership in the Post-Information Age. Parameters, Autumn.

Karonen, V. (2002): Jatkosodan kuva entistä synkempi. Helsingin Sanomat 18.5. 2002, B2.

Lash, S. (2002): Critique of Information. London, SAGE Publications.

Noskov, I. (2002): Military Ethical Reform in Eastern Europe and Russia. Presentation at International Military Ethic Symposium Riga, Latvia, 5 - 10 March.

Nygren, K.P. (2002): Emerging Technologies and Exponential Change: Implications for Army Transformation. Parameters, Summer, 86-99.

Royl, W. (2002): Euro-Military Identity in Transition. In: Toiskallio, J.; Royl, W.; Heinonen, R. E.; Halonen, P.: Cultures, Values and Future Soldiers. Helsinki, Finnish National Defence College, Department of Education, Series 2, No 11.

Shurtleff, D.K. (2002): The Effects of Technology on Our Humanity. Parameters, Summer.

Amira RAVIV, Israeli Defense Forces Colleges, Tel Aviv

TEACHING VALUES FOR MILITARY PROFESSIONALS

This article will attempt to emphasize the importance and cardinality of dealing with ethics, the complex challenge of teaching it, and the manner in which the College has designed a pedagogical-holistic concept, as a suitable instructive answer that improves itself constantly.

In attempting to demonstrate the above, two examples presenting the complex reality the IDF soldiers meet daily may be portrayed:

Situation 1: A battalion commander is carrying out operational activity in Zone A. His objective is to achieve control over a three story building, from which he is to give cover for further missions. At his disposal are infantry troops backed up by a tank. As he reaches the building he confronts heavy enemy fire coming from the 3rd floor. In parallel he hears voices of panicked women and children coming from the 1st floor. Fire exchanges continue for some time, and he understands that the civilians are trapped inside a building that is his objective. The enemy fire endangers his troops, but he must take control over the building in order to accomplish his mission. He is authorized to deploy the tank in order to do so. What dilemmas does the commander face? What considerations does the commander apply in his attempt to resolve the situation? What is the organization's expectation from the commander in such a situation?

Situation 2: A company commander is organizing a road block. He has received information concerning a terrorist organization's intention of smuggling an explosive device through a block in his section. An ambulance comes rushing to his road block with a woman about to deliver her baby, and the ambulance must urgently reach a nearby Israeli hospital. The woman and her husband are yelling and warning that they can not afford to waste time over a road block inspection, since her condition is hazardous. The battalion commander can not be reached by radio, but HQ announces that the order is to thoroughly inspect and search every vehicle (a matter of approximately half an hour). Here too, the same questions should be asked.

The answers to these events are not trivial. They both share questions of a commanders' responsibility for soldiers' lives, the completion of the mission, the maintaining of human dignity, the daily routine of Palestinian civilians, and the keeping of Israeli civilians' security. Almost every choice encloses harm in one or more dimensions. This unsecured reality, together with changes in the society's values concerning canonic organizations, requires brave treatment of morals, ethics, and leadership questions pertaining to war and armed conflict.

Since the Colleges mission is to qualify senior officers for command and staff assignments, it is our understanding that permanent treatment of ethical and normative questions is the essential backbone of any military in a democratic country, and thus, should be in every curriculum of ours. This is also because of the understanding that the development of moral doctrine is the basis of a differential identity, which transmits additional validity and depth to the military profession.

For this need our college has adopted Ethics Professor Asa Kasher's concept, claiming that every profession consists of five elements: Unique knowledge, unique skills, continuous development mechanisms, local understanding, and global understanding of the profession, i.e., ethics.

Professionalism means knowing things you did not know earlier to their depths, to develop unique skills, and to dedicate attention to implicit dimensions of the profession.

Further assumptions we adopted were the understanding that we must design an instructive plan to treat three main entities with which the military professional interacts: The *people* - the basic body for which he works, i.e. the democratic state and all its citizens; the *colleagues* - commanders, subordinates, the unit; and the *enemy*.

The context of dealing with the enemy raises the unique and new difficulty that modern 21st century militaries are confronted with - no more conventional enemies (i.e. military), but terrorist entities which conceal themselves amongst civilian population with the purpose of harming this civilian population.

Another guiding implication is that the grasping of the subject contains supplementary contradictions: a combination of abstract ideas which are connected to the military's duty in a democratic country, but also the portrayal of practical principles that provide cognitive tools and are part of the professional practice. Another contradiction shows the need to give simple and practical answers in spite of the complex and dynamic world in which officers will meet new unfamiliar situations. However, in an ever-changing reality we face an ever-growing difficulty to provide the officer with a system of orders and procedures that will give him the answers to resolve any problem that might occur. To this reality, professionals must be supplied with cognitive and analytic tools that are broad and smart, and not technical rules of *"do's and don'ts."*

Rules like these were depicted in *"IDF's spirit"* through the core values and action norms. But these demand constant clarification and inexhaustible elucidation by the individual and by the system.

In order to build an instructive skeleton we had to emerge of the understanding and the deciphering of approaches dealing with education of values and approaches dealing with learning.

As for education of values, it was possible to lean on philosophical and sociological approaches. Philosophy struggles with the ability and right of the society to educate the individuals who construct it. The question of autonomy, authenticity and self-actualization of the individual is a cognitive framework, much like the question of education in a democratic, pluralistic and multi-cultural society.

Sociology deals with education for values out of the research of social role and organizational structure of educational institutions. In their construct, their authority relationship, and their instructors teach tacit values, which are usually mainstream in the society. It is clear that the role of the educational system is to encourage positive or negative attitudes towards a desired value or a norm and to create internalization and identification processes in varied ways.

Therefore, as the famous Israeli novelist, S. Yizhar depicted, there are practically four attitudes toward education for values:

Education of values is needed and possible;
Education of values is needed but not possible;
Education of values is possible but not needed;
Education of values is not needed and not possible.

While Yizhar sided with the last attitude, the IDF and the Colleges sided with the first one.

But the phrase *"Education of values"* is quite naïve and somewhat presumptuous. It would be better to replace it with *"nurturing"* or *"reinforcement."* That is because the students of the courses arrive equipped with a professional normative system, and are not to be treated as novices or inexperienced individuals.

As mature learners, with ranks of Majors or Lieutenant Colonels, the trainees are willing to influence the learning aims, are capable of self-guidance, and are normally active. They are instrumentally interested in the teaching topics, i.e. purpose oriented. Therefore, the topics should be relevant to their everyday problems, and connected to the role of self-realization.

Finally, we are challenged with classes that are heterogeneous in their roles, combat experience, age, and perspectives. Therefore, if in any case the characteristics of the teaching methods for adult population are based on interactivity in the teaching process, utilizing variable instructing methods and the creation of an atmosphere of openness and criticism, then the treatment of values must comply with these standards as well.

The process of influencing attitudes, and the process of touching the ethic profile of a professional, is best routed by inclusive ideological education. Education of values does not operate internalization processes if the student is passively listening to a lecture. Also, a discussion being held in a hypocritical manner would influence only habits of speech and expression. The way that creates significant change is the discussion, dialogue, sharing of opinions, persuasion, and experimentation. It is obvious that in order to acquire a norm, one must first acquire knowledge, acquire facts and theories, and understand the meaning of principles, and possess the cognitive and analytical skills that are connected to that reasoning.

Even so, one must understand that agreement with certain values does not necessarily guarantee identification with them. Contradictions evolve when the individual stands across a system of contradictory pressures, and the uprising of new and difficult dilemmas. For instance, contradictions between the purity of the weapon and human life, between the responsibility for one's soldiers and the responsibility for civilian population caught up in the fire, between adhering the mission and one's personal beliefs, etc. The constant debate between cognition and feeling, the motive and the social expectation in a process of a brave group dialogue are the true backbone of the improvement of the understanding of ethics.

Since the process of learning is system oriented, and since military ethics is a comprehensive layer that touches all aspects of the commander's work, the instructors' staff has a crucial role in the process.

This process is not the realm of an academic professor, but of the instructors' and commanders' staff, as it is in the combat units. The commander is the subject of identification, is an example for soldiers, and creates stimuli and learning experience. His or her role is to provide an ethical interpretation at every opportunity. Not only in the ordered chapter in the program, but in the analysis of military topics, combat drills, military history, etc. It is imperative that the staff itself has a thorough conceptual framework for ethical evaluation, and the skill to encourage reflective thought. Therefore, we are strict about staff preparations concerning the ethical area for every course in the College, and about timely discussions in the matter. The staff itself must professionalize and familiarize itself with the consideration behind the *"IDF spirit,"* the essence of the military in the democratic state, laws of war, etc. The staff should take advantage of different events in the course, and design them into a critical ethical event, which creates an emotional stimulus to strengthen the learning.

But that is not enough. In the colleges we understood that we are demanded to think of the way we manage ourselves as an organization. Therefore we developed the College Code of conduct, which is derived from the *"IDF Spirit."* This code deals with values of learning that are connected to our existence as a military organization, to our unique norms as a training system of senior officers, and to our interactions with our environment. Hence, our code includes values such as openness, initiative, curiosity, scrutiny, collegial relations, etc.

This code is unified for staff and trainees alike, and not accidentally. It upholds standards that require everyday treatment of our ethical character, as well as the trainee's.

All this derives an integrative curriculum, which includes ethical dealing as a separate or intertwined chapter, as a formal lesson, and also as a feedback and a systematic appraisal of students, as the subject of staff's teaching, of outside lecturers, as an area through which IDF's activity is to be inspected today and in past wars, and as a criterion by which we appraise the staff, the course, or the individual. We will merge an ethical prism when discussing combat, but also when talking about the force-building routine. We shall deal with positive as well as negative events (as unfortunately these occur as well), with courage and with no favoritism. Along the way we will check whether we accomplish the goals we have set for ourselves, and what is later to be improved.

Ethics is a language and a concept. There are no boundaries to its development and perfection.

Finally, it seems that dealing with ethics is compared to learning music, which enables one to hear sounds that he has never heard before. Another allegory that can be adopted is the compass and the navigator, as once mentioned by Prof. Kasher. It is not sufficient that the compass exists as a technology. In order for it to influence his life and his professionalism, the navigator must know the compass, understand it, be convinced of its value, and know how to use it, internalize its ability, and get used to utilizing it properly.

The new nature of wars we confront, as like other western militaries, will oblige us all to find our compass that shall combine the existing with the new. A synergy of ethical insights will assist us to design substantial professional knowledge that will allow us to cope with these challenges.

Simon P. GUTKNECHT, Military Academy at the ETH Zurich, Zurich

THE INFLUENCE OF SOCIAL AND COGNITIVE SKILLS IN CHOOSING PROFESSIONAL MILITARY OFFICERS

Social skills are of primary concern in the selection of professional officers at the Assessment Center for future professional officers. Cognitive skills assessed in achievement tests play a secondary role. With regard to a study (Gutknecht, 2001), in which there was a correlation between cognitive tests as a predictor concerning success in studies, an important question has been raised: To what degree does the weight given to the assessment process justify the acceptance in the officer's program? What follows will be the partial results of the aforementioned study and further evaluations.

Assessment Center for Future Professional Military Officers

The Department for Military Psychology and Military Pedagogy is not only responsible for teaching and research in this area but is also involved in the active selection of future professional Swiss Military Officers. This occurs through the arrangement, the revision, and the implementation of the Assessment Center for future professional officers.

The term *"Professional Military Officer"* involves different capabilities. These include teaching, training, and leadership. This means that in addition to academic or general cognitive capacities, social skills are also required. That is why the method followed in the Assessment Center is designed to appraise *"social behavior"* as well as *"personal traits and features."*

Additional cognitive tests by Saville & Holdsworth Ltd (SHL) that test verbal, numerical, and figurative reasoning are also carried out during the assessment. There are no consequences for failing these tests, if the candidate's achievements in the other exercises are sufficient. A priori, it is assumed that the influence of these tests on vocational success is irrelevant. But according to studies and meta-analyses this assumption might be wrong. (Schmidt & Hunter, 1981; 1998; Kline, 2000) It is clear that these cognitive skills have a decided influence on one's success in studies. (Hunter & Hunter, 1984; Schmidt-Atzert & Deter, 1993; Schmidt & Hunter, 1998)

It is also clear that the Assessment Center has a decisive influence on the career of a professional military officer. The problem, however, is that the prognostic validity is unknown. The coefficients determined by meta-analyses vary greatly. Thus there is a need from time to time to check the reliability and validity of the

Assessment Center for future professional officers. (Annen, 1995; Gutknecht, 1998; Wey, 2002)

In a recent graduate paper an evaluation was made to determine whether or not the high degree of importance in the selection process given to the Assessment Center was justified. (Gutknecht, 2001)

Because of the lack of certain data it was necessary to divide the evaluation into two separate studies. One evaluated *"study success,"* and the latter (which is not considered in this paper) *"vocational success."* Cognitive tests have only been added since 1996. This means that there is a dearth of information concerning *"vocational success"* with regard to candidates who had already taken these tests. Because it was assumed that these cognitive tests influenced the study success, they were taken into consideration in a separate model, *"study success."*

"Study success" was validated through a sample of 72 subjects in this graduate paper. The latent variable *"social skills,"* which was obtained from the estimates made by the Assessment Center, served as a predictor. The results of cognitive tests are called *"cognitive skills"* and school achievement is called *"school performance."* The criterion variable was *"study success"* and was operationalized by performances in the first year. The entire procedure was evaluated through the method of *"structural equation modeling,"* which allows calculations for latent variables. This method is a mixture of multiple regression analysis, path modeling, and confirmatory factor analysis. (Everitt, 1996) This procedure allows one to determine construct and prognostic validity. Analysis is done with the statistics program AMOS 4.0. (Arbuckle & Wothke, 1995)

The measurement models of each latent construct were determined to be accurate. This means they were constructionally valid. Also, the *"Fit"* of the overall model seems to be good (Fit indices are illustrated in Figure 1).

As expected, according to the hypothesis (see Figure1), the variables *"school performance"* and *"cognitive skills"* were significant predictors related to the criterion. The coefficient of the former is .38 and of the latter .31, which means the tests predict additional variance in the variable *"study success"* (incremental validity). As expected, influence of the variable *"social skills"* was weak and not significant (it only accounted for 6% of the variance).

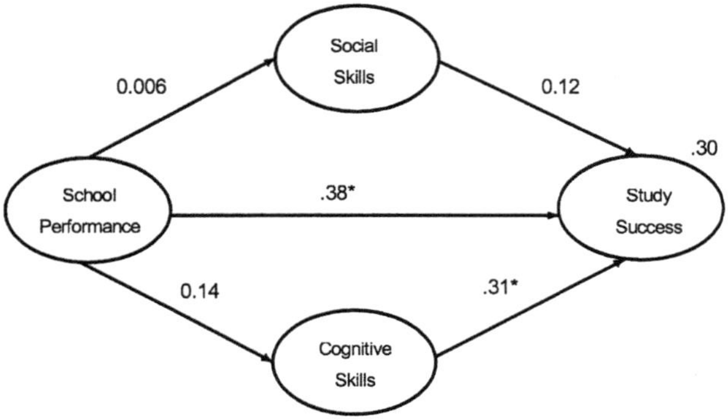

* = p<.05

Figure1: Study Success (TLI (Tucker Lewis Index) = .906; RMRS (root-mean squared-error-of-approximation) = .052

As mentioned above, *cognitive skills* are considered to be a significant determinant of *vocational performance*. This is also possibly true in the selection process of professional military officers. According to the 2001 study the variable *study success* has a major effect on the variable *vocational performance*. It is reasonable to assume that the *cognitive skills* measured by the cognitive tests also influence the variable *vocational success* of professional officers. This assumption is supported by results generated by meta-analysis. (Schmidt & Hunter, 1981; 1998)

Because of the lack of *vocational success* data, it will be necessary to carry out elaborated studies to test this hypothesis which will only be possible in four to six years.

In the follow-up analyses the previous findings (without taking school performance into account) are replicated to investigate and find out if there are any additional correlations between *cognitive skills* and *social skills* as measured in the ACABO.

This is important because there are studies that underline the influence of *cognitive skills* on Assessment Center success. (Scholz & Schuler, 1993)

The sample consisted of n=165. The results are illustrated in Figure 2.

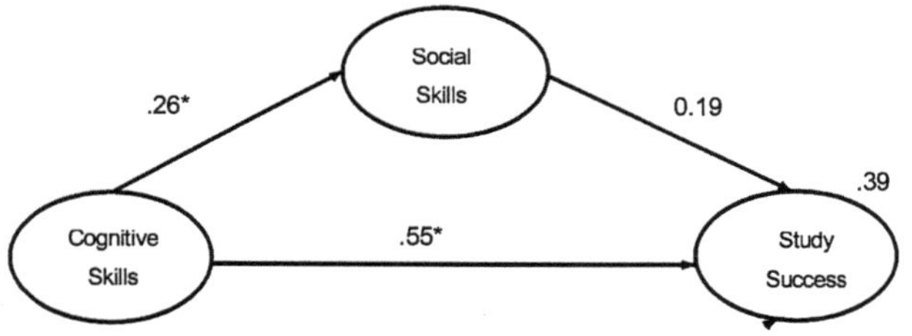

Figure 2: Study Success TLI (Tucker Lewis Index) = .999; RMRS (root-mean squared-error-of-approximation) = .023

The findings of the 2001 study were replicated. The variable *cognitive skills* has a major influence on the variable *study success* (55*), as well as on the success in the Assessment Center (*social skills*). The latter finding is interesting and is a further confirmation of the results of Scholz und Schuler (1993). However, one of the main questions remains: How do we make a reasonable difference between *social skills* as operationalized by the Assessment Center tests and *cognitive skills*? These general constructs may have a lot in common. These results will have to be reanalyzed and replicated in a larger context. Further studies are planned in order to answer these questions.

Final Remarks

In conclusion, one important thing can be said: The *cognitive skills* as measured by tests in the Assessment Center for future professional officers play a major role in predicting the individual's *study success*. The results were replicated in a separate study.

But in order to determine the value of any instrument it is necessary to take into consideration the selection process as a whole. In the research paper the *construct* validity and the *prognostic* validity of each instrument within the selection process must be tested. It is important to be able to compare the individual instruments with each other and to determine whether or not there is incremental validity. It is the incremental validity that is most important when attempting to determine the value of an additional instrument. (Cook, 1998)

Future studies have to take into consideration all possible predictors within the entire selection process. In order to effectively determine the influence of the ACABO and its cognitive tests on *vocational success,* a more sophisticated test design will be needed.

In the opinion of Höft *"the prognostic and economic efficacy is not only a question of quality control but is also a justification of one's psychological work with personnel."* (Höft, 2001: 618)

References

Annen, H. (1995): Konstrukt- und kriterienbezogene Validität des MFS-Assessment-Centers. Unveröffentlichte Lizentiatsarbeit an der Philosophischen Fakultät I der Universität Zürich.

Arbuckle, J.L. & Wothke, W. (1995): Amos 4.0 User's Guide.Chicago, Smallwaters.

Cook, M. (1998): Personnel Selection: Adding Value through People. 3rd ed., Chichester, Wiley.

Everitt, B.S. (1996): Making Sense of Statistics in Psychology - A Second Level Course. Oxford, Oxford University Press.

Gutknecht, S.P. (1998): Beobachterübereinstimmung im ACABO: Ein Vergleich der dimensionsspezifischen Übereinstimmung zwischen Milizbeobachtern und Instruktoren sowie einige generelle Befunde zum Beurteilungsverhalten der Assessoren. Unveröffentlichte Praktikumsarbeit, Institut für Psychologie, Universität Bern.

Gutknecht, S.P. (2001): Eine Evaluationsstudie über die verschiedenen Instrumente der Berufsoffiziersselektion und deren Beitrag zur Vorhersage des Studien- und Berufserfolges. Unveröffentlichte Lizentiatsarbeit am Institut für Psychologie, Universität Bern.

Höft, S. (2001): Erfolgsüberprüfung personalpsychologischer Arbeit. In: H. Schuler (ed.): Lehrbuch der Personalpsychologie, Göttingen, Hogrefe, 618-648.

Hunter, J.E. & Hunter, R.F. (1984): Validity and Utility of Alternative Predictors of Job Performance. Psychological Bulletin, 96, 72-98.

Kline, P. (2000): The Handbook of Psychological Testing. 2nd Ed., London, Routledge.

Schmidt-Atzert, L. & Deter, B. (1993): Intelligenz und Ausbildungserfolg: Eine Untersuchung zur prognostischen Validität des I-S-T 70. Zeitschrift für Arbeits- und Organisationspsychologie, 37, 52-63.

Schmidt, F.L. & Hunter, J.E. (1981): Employment testing: Old theories and new research findings. American Psychologist, 36, 1128-1137.

Schmidt, F.L. & Hunter, J.E. (1998): The Validity and Utility of Selection Methods in Personnel, Psychology: Practical and Theoretical Implication of 85 Years of Research Findings. Psychological Bulletin, 124, 262-274.

Scholz, G. & Schuler, H. (1993): Das nomologische Netzwerk des Assessment Centers: eine Metaanalyse. Zeitschrift für Arbeits- und Organisationspsychologie, 37, 73-85.

Schuler, H. (1998): Psychologische Personalauswahl - Einführung in die Berufseignungsdiagnostik. 2. unveränderte Auflage, Göttingen, Verlag für Angewandte Psychologie.

Wey, M. (2002): Assessment Center für angehende Berufsoffiziere: Eine Analyse der Interrater-Reliabilität in Bezug auf die Gesamtbeurteilung sowie die dimensions- sowie übungsspezifischenUrteile. Unveröffentlichte Praktikumsarbeit, Militärakademie an ETH Zürich.

Guenther FLECK, National Defense Academy, Vienna

Teaching Ethics: A Psychological Perspective

Introduction

There is no doubt that most people around the world know to some extent the difference between good and evil. It appears to be true that cultural influences have a strong impact on individual differences concerning value development and value orientation. Unfortunately, myriads of situations all across the globe demonstrate that many people find it difficult or are unwilling to act according to ethical rules. This becomes exceedingly evident under conditions of war and in social, religious, political, and ethnic controversies often resulting in violent conflict. The phenomenon of unethical behavior, ranging from more or less mild offences against moral rules to extreme war crimes, also appears during military operations and is sometimes even conducted by regular soldiers and established military forces.

Governments and military establishments try to avoid the appearance of such misbehavior by introducing adequate rules. But it is evident that regulations by rules (e.g. laws) alone are insufficient to generate ethical behavior. Consequently, all civilized societies undertake a significant effort to teach their citizens and soldiers the basics of ethics and provide a framework for moral decision-making and behavior.

Nevertheless, teaching ethics is often done in a merely formal and rational way without taking the emotional side of the students into account. But this emotional side has to be affected also if teaching ethics is supposed to show successful results. This paper provides a heuristic model which highlights some important aspects of teaching ethics from the perspective of individual psychology.

Ethical Behavior: The Individual Perspective

If we ask for the conditions of ethical behavior from an individual perspective, then it seems reasonable that the following basic requirements have to be fulfilled:

(I) Being able to recognize unethical behavior. The individual has to be able to recognize if something is going wrong with respect to ethics. This implies that a system of values has been acquired, and that these values can be used to assess situations adequately. In many cases cultural differences may have a strong impact on moral decision-making and a great deal depends on the intercultural competence of the observer to make the right conclusions. (II) Knowing the rules of how to correctly behave ethically. This may often lead to the dilemma that someone is recognizing unethical behavior, but at the same time does not know what to do. (III) Capability to act according to the rules in the correct way, which means that the individual should generate the necessary behavior for moral action. (IV) Motivation to act in the correct way. The individual has to be motivated (i.e., must feel the wish) to act in the correct way. Whatever the reason may be, people are often unwilling to act according to ethical rules. (V) Situational awareness with respect to ethical behavior. To sustain situational awareness with respect to ethical behavior as a kind of general guideline is particularly important in military operations and on the battle field. Situational awareness should function as a kind of constant and automatic scanner to detect problematic situations and to intervene effectively in time.

While by way of socialization and education most people have developed the necessary skills with respect to ethical behavior, many often manifest contradictory actions in different situations. Why do people act against their better insight and knowledge? If one does not believe the cause is due to an aggressive drive or other instincts, other alternative explanatory concepts for unethical behavior have to be considered. Since people normally have the freedom to decide what to do, a human information processing approach seems to be more explanatory. Although daily human behavior is usually produced and controlled automatically, human beings tend to think before they act. The potential to anticipate the outcomes of different actions provides the individual with responsibility for his or her actions. Thus, it appears as if subjective responsibility has to be understood as the core element of ethical behavior.

Personality, Human Information Processing, and Ethical Behavior

The model presented in the following is based on the interactional view of personality, which reduces individual experience and behavior neither to isolated personality traits nor to single situational demands. Moreover, individual behavior is better understood as a result of dynamic interactions between personality variables and environmental influences.

Endler & Magnusson provide a characterization concerning an interactional psychology of personality. They consider the individual as an intentional active agent in an interactive process. Actual behavior is a function of a continuous process or multidirectional interaction (feedback) between the individual and the situation that he or she encounters. On the person side of the interaction, cognitive factors are the essential determinants of behavior, although emotional factors do play a role. On the situation side, the psychological meaning of the situation for the individual is the important factor. (Endler & Magnusson, 1976: 968)

Ever since the Endler & Magnusson study of 1976 emotional factors have been considered to be increasingly important. Human information processes are now mostly viewed as interwoven affective-cognitive processes. In Marcel's words:

> *"Traditional approaches in psychology have separated cognition, conation, and affect. So-called cognitive psychology mostly treats our cognition as cold or affectively neutral. Yet one thing I know about my day-to-day consciousness is that it is almost never neutral. Everything I do, think, perceive has an affective tone, positive or negative and of a particular kind; furthermore, I have an attitude to everything I do, think, and perceive."* (Marcel, 1988: 173)

Meanwhile numerous approaches do exist concerning the affective-cognitive dimension of behavior. Toda summarizes the role of emotions with regard to decision-making by pointing out that instead of dismissing emotions as irrational agents, they should be understood in their nature and how they influence the decision-making process in order to acquire better control of them. (Toda (1980: 133)

His findings demonstrate that the whole set of *emotions* forms a system that is developed over time and generically programmed. The set of emotions serves the purpose of making decisions in an environment that can be characterized as primitive and wild. The *non-emotional*, more analytical decision system is a product of a much later period of evolution which serves along with other higher cognitive-analytical functions primarily to supplement, but not to replace the emotion system. Even though these two systems are often in conflict, the cognitive decision system does not operate without the help of the emotion system; without desires, loves, and hates there hardly would be utilities.

In sum, treating ethical behavior from a modern information processing perspective has to include the affective side of the human being. But also, the assumption opens the possibility of studying the emotion system as a purposeful, rational decision system in its own right.

Mind Sets and Ethics

Throughout the course of their lives human beings create world views which allow them to order things and provide subjective stability and security. The individual world view also includes a cultural representation system distinguishing between *good* and *bad*, which for the purpose of this paper is termed the *"ethical mind set."* The creation of ethical mind sets is embedded in the process of internalization, a key concept in understanding the acquisition of cultural representations.

By referring to Spiro (1987), D'Andrade explains the four different levels of internalization. On the first and lowest level, the individual gets acquainted with various representations of the cultural system without, however, assenting to its normative foundation. At the second level, cultural images are obtained as clichés. Spiro gives the example of people believing that Jesus died for their sins, but have no sense of sin. On the next level individuals hold their beliefs to be true, correct, and right. Cultural representations structure the behavioral environment of social actors and guide their actions. At this stage, individuals have sense of sin, and cultural images are internalized. At the final level, *"the system of cultural representations is not only internalized, it is highly salient. The cultural system not only guides but instigates action, and the entire system is invested with emotion. Thus, believing that Jesus died for man's sins, the believer is filled with anxiety about his own sins, and driven to try to atone for these sins in prayer and deeds, and filled with relief and joy at evidence of being saved."* (D'Andrade, 1995: 228)

Based on these four steps of internalization three types of ethical mind sets are proposed: Type I represents the deep structure type; type II the surface structure type, and type III the non-structure type.

The *deep structure* type is characterized by his or her highly-internalized value and belief systems. People representing this mind set maintain their value orientation in most of all situations. The *surface structure* type knows about the *correct* values and how to behave according to them, but has not internalized his or her value and belief systems in a deep way. This type may behave very differently with respect to situational changes and sometimes act extremely unethically. Type III is the *non-structure* type that stresses high emotional deficits in interpersonal relationships. Although these people know about ethics (i.e. decision-making between good versus bad actions) they are not willing to behave according to them. They are simply not interested in ethics and morality.

The *non-structure* character can be interpreted as a complex personality disorder, the so-called psychopath, manifesting different psycho-pathological key symptoms. These are in the emotional interpersonal domain characteristics such as glibness and superficiality, egocentricity and grandiosity, lack of remorse or guilt, lack of empathy, deceitful and manipulative interaction styles, and shallow emotionality; in the social domain impulsivity, poor behavior controls, need for excitement, lack of responsibility, early behavior problems, and adult antisocial behavior. (Hare, 1999)

Conditions Affecting Unethical Behavior

Far from being complete, three important conditions affecting unethical behavior shall be briefly discussed: deindividuation, selective identification, and alterations in thinking under war conditions.

It is hypothesized that individuals from the *surface structure* type and *non-structure* type are prone to act unethically under these conditions, with the main difference being self-control. While *non-structure* type individuals have deficits of self-control even under normal conditions, *surface structure* type individuals are largely able to manifest and maintain self-control under normal conditions.

The first condition, deindividuation, understood as a psychological state, is a social phenomenon which is aroused when individuals join crowds or large groups. Characteristics of the state of deindividuation are a diminished awareness of the individual self and reductions in self-restraint and normative regulations of behavior. (Diener, 1980; Zimbardo, 1969) This may lead to different variations of collective behavior of violent crowds (e.g. the lynch mob).

Selective identification, on the other hand, implies that people like to identify with other persons who share certain common interests or characteristics with them. (cf. Tart, 1975) The great danger of selective identification lies in the possible rejection of those persons who belong to other races, political parties, nations, religious groups, scientific communities, language populations, and so on. Under certain circumstances these *"other people"* may be labeled as enemies, and may be de-humanized as a final consequence. The result, then, is that other human beings are treated as non-human beings, often as beasts, which subjectively justifies treating them inhumanely.

A third condition encouraging unethical behavior is the alteration of thought patterns, which can be observed frequently in wartime. Aspects of thinking in peacetime (P) versus thinking in wartime (W) have been elaborated by LeShan, some of which are presented below: (LeShan, 2002: 35-36)

[I] *(P): Good and evil have many shades of gray. Many groups with different ideas and opinions are legitimate. Their opinions, and things in general, are relatively good or bad, satisfactory or unsatisfactory, stupid or intelligent.*

(W): Good and evil are reduced to Us and Them. There are no innocent bystanders; there are only those for or those against us. The crucial issues of the world are divided into black and white. Opinions on these matters are absolutely right or absolutely wrong.

[II] *(P): Now is pretty much like other times. There are more of some things, less of others, but the differences are quantitative.*

(W): Now is special, qualitatively different from all other times. Everything is cast in the balance; whoever wins now wins forever. It is the time of the final battle between good and evil - of Armageddon, of Ragnarok, of The War to End All Wars.

[III] *(P): The great forces of nature, such as God or human evolution, are not particularly involved in our disputes.*

W: "Gott mit uns," "Manifest Destiny," "Dieu et mon Droit," "History fights on our side," and other such slogans indicate our belief that the great motivating forces of the cosmos are for Us.

[IV] *(P): When this present period is over, things will go on pretty much as they have in the past.*

(W): When this war is over, everything will be vastly different. If we win, it will be much better; if we lose, terribly worse. The world will be deeply changed by what we do here. Winning or losing will change the meaning of the past and the shape of the future.

[V] *(P): We can talk with whom we disagree. Negotiation is possible.*

(W): Since the enemy is evil, he naturally lies. Communication is not possible. Only force can settle the issue. We tell the truth (news, education). They lie (propaganda).

[VI] *(P): All people are fundamentally the same. Differences are quantitative.*

(W): "We" and "They" are qualitatively different, so different that the same actions are "good" when we do them, and "evil" when the enemy does them. There is doubt that we and they really belong to the same species.

It is self-evident that such alterations in thinking may function easily as triggers for unethical behavior.

Ways of Teaching Ethics

Basically, two different modes of teaching ethics can be distinguished: Teaching ethics via *rational argumentation* (third person perspective) and teaching ethics via *sensitizing* (first person perspective).

The first mode represents the classical way of teaching by focusing on well-elaborated philosophical concepts. The second mode represents a newer way of teaching by focusing on the student's emotional experience. In the first mode students learn something about ethics; in the second they become familiar with ethics.

Teaching ethics from an experiential point of view has to consider the constraints of classical teaching and to ask for effective alternatives. The question is how to overcome the constraints. Two main approaches may be considered with regard to enhancing the individual ethical strength: The improvement of those personality variables that determine individual ethical behavior on the one hand, and the sensitization of individuals for other people's vulnerabilities on the other.

Developing Ethical Skills via Systematic Self-Reflection

Modern human sciences, especially psychology, offer a variety of methods to develop and enhance ethical skills such as personal responsibility, empathy, tolerance, techniques to cope with one's own aggressive impulses (e.g. craving for revenge), and to cope with the temptation of dehumanizing adversaries and enemies. A method of systematic self-reflection as a basis for personal growth may be used as an *"antidote"* against unethical behavior. (Fleck, 1999)

Bühler and Ekstein (1973) claim that few people have learned to think life through, and to gain deeper knowledge about the processes of life. Only when pressure of time or problems have become predominant or a crisis has caused suffering, people start to think and reflect on important events and areas of life. While doing so, reflection is often practiced in an undifferentiated and self-restricting manner, which may enhance stress, decrease subjective well-being, and may finally lead to chronic disease.

The method of systematic self-reflection offers a special way to tackle important events and areas of life systematically, characterized by intentionality and regularity. The rationale of systematic self-reflection is based on three elements: On defining *goals* ("Wherefore should one reflect?"), on selecting *contents* ("About what should one reflect?") and its *formal aspects* ("How should one reflect?"). The main purpose of this method is seen in the acquisition of self-knowledge and self-complexity since empirical evidence shows that people with more self-complexity are less susceptible to stress than people with less self-complexity (Linville, 1987).

The first step toward systematic self-reflection is to become aware about the *goals* that one wants to reach. This depends on subjective preferences, values, and actual needs. It is not possible to give a complete list of all goals, but the following ones may have high relevance for most people: self-congruence, defining one's position in life, clarifying one's own roles, solving personal problems, removing self-deception, self-actualization, developing a life plan, enhancing self-knowledge, and maintaining a good health state. The leading question will focus on the things which are really important in life.

The search for goals directly leads to the *contents*. Clinical experience has proven that contents feature different areas of life, such as profession, partnership, family, religion, politics, arts, life events, world view, value orientation, thinking habits, status of health, and sources of stress. (Kreitler, 1999)

As far as the *formal aspect* is concerned, there are two basic approaches toward systematic self-reflection. The first one may be called rational self-reflection. There, the contents are processed by means of logical-analytical methods including internal dialogue and mental imagery. Via this method, for example, one can learn to make latent belief systems explicit and question them critically, thereby gaining new insights and perspectives.

A good way to tackle the various goals and contents is offered by training in the so-called consciousness disciplines, a term covering various methods (e.g. meditation, self-hypnosis, lucid dreaming) to expand ordinary consciousness. (e.g. Deikman, 1983; Walsh, 1980; West, 1987; Wolman & Ullman, 1986) One important goal of these methods is to gain new insights into the process of reality construction. (DelMonte, 1987) In the field of science Tart created the concept of state-specific sciences, arguing that altered states of consciousness (compared to our normal waking state) may offer possibilities of totally different views of phenomena and data interpretation. The consciousness disciplines allow the individual to learn how to dis-identify from the contents of consciousness (e.g. thoughts, emotions, desires). This may help to produce an inner distance from the contents of consciousness, temporarily making it easy to explore alternative perspectives in scientific reasoning. (Tart, 1972; 1975; 1998: 224)

Sensitizing Individuals to Other People's Vulnerabilities and Dignities

Since this approach of teaching ethics is elaborated and discussed in detail by Rainer Born in this volume, only a short comment is necessary. Central to sensitizing individuals to other people's vulnerabilities and dignities is the application of the first person perspective. This implies that the means of teaching is not the rational argumentation about other people, but rather the identification with other people. Induced identification with other people may lead to a better understanding of others and allow empathizing with the *"other."*

All these methods contribute well to individuals' personal growth, which represents the most important *"antidote"* against immorality.

Summary and Conclusion

In this paper a framework was offered concerning some principles regarding teaching ethics. It was argued that well-elaborated philosophical concepts represent a necessary but insufficient condition in ethical education: Teaching ethics cannot be concluded via rational argumentation alone and should therefore be extended to the inclusion of emotional conditions.

After considering some basic conditions for individual ethical behavior the three individual forms of ethical orientation were introduced and some situational conditions triggering unethical behavior were outlined. The analysis made clear that especially people of surface structure type mindsets have to be taught in ethics by means of affecting their emotionality in order to make them more sensitive.

Whereas the deep structure type of person may further improve his or her already well-developed value orientation including sympathy for other people through ethical teachings, there is, as far as scientific studies show, no effective approach to sensitize the non-structure type with regard to ethics.

While ways to influence the non-structure type will be the subject of further research in psychology, it is assumed that teachers and instructors in military ethics have to deal with individuals whose personalities are susceptible to the conveyance of moral values and principles.

References

Bühler, Ch. & Ekstein, R. (1973): Anthropologische Resultate aus biographischer Forschung. In: Gadamer H.-G. & Vogler P. (eds.): Psychologische Anthropologie. Stuttgart, Thieme.

D'Andrade, R.G. (1995): The development of cognitive psychology. Cambridge, University of Cambridge Press.

Deikman, A. (1982): The observing self. Boston, Beacon.

Diener, E. (1980): Deindividuation: The absence of self-awareness and self-regulation in group members. In: Paulus, P. B. (ed.): The psychology of group influence. Hillsdale, NJ; Lawrence Erlbaum, 209-242.

DelMonte, M.M. (1987). Constructivist view of meditation. American Journal of Psychotherapy, 41, 286-298.

Endler, N.S. & Magnusson, D. (1976): Toward an interactional psychology of personality. Psychological Bulletin 83, 956-947.

Fleck, G. (1999): Self-reflection and de-reflection: Two modes of awareness guiding individual experience and action. Cognition, Brain, Behavior 3, 223-227.

Hare, R.D. (1999): Without consciousness. The disturbing world of the psychopaths among us. New York, Guilford.

Kreitler, S. (1999): Self-reflection and health: The cognitive orientation approach. Cognition, Brain, Behavior 3, 229-238.

LeShan, L.L. (2002): The psychology of war: Comprehending its mystique and its madness. New York, Helios Press.

Linville, P.W. (1987): Self-complexity as a cognitive buffer against stress-related illness and depression. Journal of Personality and Social Psychology 52, 663-676.

Marcel, A.J. (1988): Electrophysiology and meaning in cognitive science and dynamic psychology - Comments on *"unconscious conflict. - A convergent psychodynamic and electrophysiological approach"*. In: Horowitz, M.H. (ed.): Psychodynamics and cognition. Chicago, University of Chicago Press, 169-189.

Spiro, M.E. (1987): Collective representation and mental representation in religious symbol systems. In: Kilborn, B. & Langness, L. (eds.): Culture and human nature. Theoretical papers of Melford E. Spiro. Chicago, University of Chicago Press.

Tart, C.T. (1972): States of consciousness and state-specific sciences. Science, 176, 1203-1210.

Tart, C.T. (1975): States of consciousness. New York, Dutton.

Tart, C.T. (1998): Investigating altered states of consciousness on their own terms. A proposal for the creation of state-specific sciences. Ciencia e Cultura (Journal of the Brazilian Association for the Advancement of Science) 50, 2/3.

Toda, M. (1980): Emotion and decision making. Acta Psychologica 45, 133-155.

Walsh, R. (1980): The consciousness disciplines and the behavioral sciences: Questions of comparison and assessment. American Journal of Psychiatry 137, 663-673.

Zimbardo, P.G. (1969): The human choice: Individuation, reason, and order vs. deindividuation, impulse, and chaos. In: Arnold, W. J. & Levine, D. (eds.): Nebraska Symposium on Motivation. Vol. 17, Lincoln, NE, University of Nebraska Press, 237-307.

Human Perspectives of Applied Ethics

Jarmo TOISKALLIO, National Defence College, Helsinki

INDIVIDUALS, ETHICS, AND MILITARY PEDAGOGY

> *"As technology continues to evolve, so too does our operational doctrine... and the warrior ethic of the infantryman remains steadfast."* (Honoré & Cerjan, 2003)
>
> *"The future soldier will be highly trained and skilled ... the small size of the unit will require each man to become an expert ... Soldiers will have to have a deep sense of unit loyalty and, simultaneously, a strong sense of personal independence."* (Friedman & Friedman, 1988)

Introduction

It is a part of the contemporary military discourse to see that competent individual soldiers (thinking, creative, multi-skilled, adaptive, agile, etc.) will have a growing role in the armed forces. Accordingly, it is a challenge for military education and training that rapid and dispersed operations will require men and women who are part of a cohesive team, but at the same time they have to be capable of operating independently.[14]

The two quotations above refer to two current elements of the human dimension in the military - ethics and individual expertise. As such they are, of course, extremely important and interesting for military pedagogy (MPED). The first sentence is a part of the new US Army program *"Warrior Ethos."* The second citation is a typical phrase highlighting the difference of training between the old *"rigid way of mass armies"* and *"the model for the soldier of the future."*

Are we ready and able to train and educate expert soldiers with strong senses of personal independence and with high levels of ethics? And is there some difference between ethos and ethics? Ethos means habits, socially shared values, rules, and modes of behavior. As such I take ethos to be a step towards ethics. Habituation, growing into the ethos of a social group, is a necessary phase to become an ethical person. But on the other hand, there cannot be any deeply functioning ethos without ethics and ethical persons as members of the community.

The idea behind the program Warrior Ethos is to see the future military missions, environments, and battles as extremely violent, unpredictable, and multidimensional. Soldiers and leaders will be under great stress – physically and psychologically – no matter what their rank, specialty, or location on the battlefield. That is why individuality and ethos must be taken even more

---

[14] US Joint Vision 2020.

seriously than before. Already the very word *"warrior"* can be seen as a concept containing the theme of individuality, and especially that of ethos. (French, 2002)[15]

It is defined that the soldier of the near future has to be *"multi-skilled," "adaptive,"* an *"adaptable learner,"* and *"self-aware."* (Johnston; Leibrecht; Holder; Coffrey & Quinkert, 2002) In the same way, although a decade earlier, the new workers of the post-industrial society have been described as *"multi-skilled," "flexible,"* and *"aware."* (Casey, 1995) It is seen that a worker's ability to learn and adapt becomes more important than his or her past training. It is the individual qualities that have been strongly underscored.

For MPED, it is the principal task to inquire into the development of human capacities and competencies in the military in terms of learning and growth. More specifically, MPED is interested in the human beings in the armed forces: How to develop the action competence of soldiers as persons and agents? It is just here that we meet the profound educational-ethical challenge: How to support individuals to constitute themselves as moral subjects of their actions? This is a process of growth that can also be called becoming ethical.

An agent is an acting subject. Action is not plain behavior; it is always intentional. One acts for a reason, in order to do something, and as an expression of some values and attitudes. The term *"subject"* has the meaning of the acting person; a person as a moral subject of his/her own action. That is why the focus of this article is on the concepts of *"self"* and *"identity"* which will be connected to the whole question of ethics. Without acting subjects, all kinds of ethical rules and codes will remain dead words. And this kind of understanding of the concepts of action, agent, subject, self, and identity is my background for using the term *"action competence."*

The Task and the Framework

The aim of this paper is to investigate the theme of the individual soldier as an ethical contribution and challenge for MPED. It is ethical as far as it assumes the soldier to be or to become a person and an agent, different from the machine-type soldier that Michel Foucault so finely describes in his book Discipline and Punishment: *"...the soldier has become something that can be made; out of a formless clay, an inapt body, the machine ...."*[16] As a person and as an agent the

[15] French writes that *" 'Warrior' should not be used to describe every individual who ... fights ... a war. The term would have more strength if we reserved it to apply only to those war fighters who meet other important criteria ...more significant, than that of having taken up arms against an enemy. ... we should first ask why they fight, how they fight, what brings them honor, and what brings them shame."*

[16] Foucault describes the type of soldier that was developed by the end of the eighteenth century.

soldier cannot be described within the paradigm of *"Man-the-Machine."* Agency (to be an agent) requires *"that one understands one's relation to oneself within a context of cultural meaning."* (Jurist, 2000:127) A machine cannot understand cultural meanings and itself. It is also impossible for an abstract or plain individual outside a community and its culture. I think this is the way we should approach the question about the future individual soldier both as an expert and as a good team member.

The individual soldier as an ethical subject - as an agent - is the focus of this study. That kind of aspect comes to the fore when we say that as rational human beings, soldiers should be held responsible for their own actions.[17] (Neill, 2000) Perhaps the main challenge for MPED should be the development of soldiers as human beings who are responsible for their actions. That is the core meaning of *"agent."*

It is well known that one of the most important constituents of the human dimension in the military is morale. High morale of the armed forces is one of its priorities. That is a challenge covering the individual and unit levels as well as the armed forces as a whole. Morale is said to include the following elements: discipline, a sense of commitment, self-control, a sense of self-respect, integrity, and devotion to military duty. (Zelenkov, 2001) Morale belongs into the core of fighting power as it is said, for instance, in the NATO paper "Land Operations in the Year 2020:" *"Conflict is, and will remain, essentially a human activity in which man's virtues of judgment, discipline and courage - the moral component of fighting power - will endure."*

However, at the level of individuals this paper will speak of ethics instead of morale.[18] Ethics should be understood as a reflection on morale and ethos aiming at self-education and self-improvement. That is how Michael Foucault's phrase should be understood: *"Ethics is a practice; ethos is a manner of being."* (Foucault, 1986)

The concept of *"action competence"* will be used to refer to the agency of the soldier. *"Action competence"* searches for practicality and the intimate link between theory and practice in the footsteps of the concept of *"subject as embodied agency."*[19] It is just the idea of embodied agency that safeguards the idea of ethics from individualism. The individual soldiers are themselves responsible for the ethical choices they make, but they serve the interests of their country.

---

17 The British professor Ian Bellany says that society *"increasingly expects its soldiers to act as rational human beings and to be treated as such."*

18 For instance, the philosopher Emmanuel Levinas says that morality belongs to social and organizational spheres, but to be ethical is always something very individual and personal.

19 *"Embodied agency"* is the concept used by Charles Taylor.

It is reasonable for this kind of investigation to use the framework of philosophical anthropology. Philosophical anthropology is the attempt to elucidate the basic constitution of human subjectivity. In short, our understanding of human nature is its main area of interest. Some of the central concepts of philosophical anthropology are human agency, personhood, identity, and interpersonal recognition. (Laitinen, 2003) The philosophy of body has intimate connections with philosophical anthropology; one of its main questions concerns the role of our body in our experience of the world. (Proudfoot, 2003) For MPED this kind of holistic approach to *"embodied agency"* is extremely important, because it tries to understand the acting person, not only the mind or the functions of the brain.

In recent years there has been a growing interest in the cognitive skills of soldiers. But it does not cover the levels of ethics, embodied agency, and action competence. A good typification is presented by Chiu, Hong and Dweck in their integrative model of psychological functioning (Chiu; Hong & Dweck, 1994). They show that modern cognitive psychology mainly deals with the processes customarily named *"intelligence"* only, when the actual ethical problem solving takes place at the level customarily named *"personality."*

	Basic Process Level	**Knowledge Base Level**	**Actualized Behavior Level**
INTELLIGENCE	*Basic Mental Operations*	*Factual Knowledge, Skills*	*Academic Problem-Solving*
			Interpersonal Problem-Solving
PERSONALITY	*Basic Motivational and Affective Processes*	*Values, Beliefs, Standards*	*Moral-Ethical Problem-Solving*

Picture 1: Ethical problem solving with cognitive skills.

Towards self-improvement

Is it possible to train soldiers to become action competent in the sense of agency? What is needed is not only training, but education, self-education, and self-improvement. The process should start as a habituation, as taking the individuals as members of the ethical life of the community.[20] This is the idea of the *"embodied agency,"* Charles Taylor's notion of the human self as deeply embedded in community. This kind of view puts heavy challenges not only on the shoulders of the individuals, but also on the cultures of military

[20] *"Ethical life"* is used in the meaning of the German term *"Sittlichkeit"* based on *"Sitten"* (customs, habits). It is in the meaning of *"Sittlichkeit"* that ethics is coupled with the forms of life of the community.

organizations. As Donald Neill writes, the armed forces must take *"a long, hard look"* at traditional military culture in order to take its beneficial aspect and cut away the harmful ones. (Neill, 2000: 39)

It is not the most important challenge for MPED to teach ethical principles and moral codes. The profound question is how to become an agent - or in other words, an ethically competent soldier. A *"humanized version of ethics"* with its elements of *"moral resources"* and *"moral identity"* is needed. (Glover, 2001) It is not enough to memorize a military ethical code or the rules of warrior ethos. The *"self"* or identity of the soldier has to be formed by his/her own actions, by self-education and self-improvement.

In the program *"Warrior Ethos"* it is determined that warrior ethos is the foundation for the soldier's total commitment to victory in peace and war: *"Soldiers who live Warrior Ethos put the mission first, refuse to accept defeat, never quit and never leave behind a fellow American. They have absolute faith in themselves and their team."* In other words, the soldiers must have a strong moral identity. But it seems to include only *"we."* It does not include the potential adversaries as human beings; they are labeled only enemies who have to be killed. In his inquiry into the moral history of humanity, Jonathan Glover underscores the dark side of soldiering: often it happens in the circumstances of war that soldiers lose their moral self-control (he calls it the erosion of the moral resources) and start to carry out acts of cruelty, for instance physical and psychological torture. There can even be *"the love of cruelty"* and *"the festival of cruelty"* with massacres.

It is a profound problem if the military training is based on the idea that *"soldiers have to be brutalized and tribalized ... if they are to operate effectively."* Griffith (2000) It is just tribalism that Glover takes as one of the most important sources of cruelty. On the other hand, that is why the *"Warrior Codes"* are needed: *"Warriors need a way to distinguish what they must do out of a sense of duty from what a serial killer does for the sheer sadistic pleasure of it."* (French, 2002)

It should be the core task for MPED to further develop the model of action competence referring to the potentiality of a soldier to be an agent in complex and unforeseen situations. (Toiskallio, 2003) It is important to see the soldier *"as essentially an embodied agent, engaged with the world."* (Taylor, 1995:22) In other words, MPED should not only speak about ethical principles, but more about persons and their activities of ethics rooted in the life-world. That is why it is extremely challenging to see ethics as Foucault understood it, as self-improvement - *"the kind of relationship you ought to have with yourself."* (Hacking, 1989)

The ethical component has a crucial place in the totality of action competence, because it has an integrative and executive role in decision making: how should I act in order to act in a right and good way? When reflecting the pedagogical principles for the development of ethical action competence, it seems fruitful to apply the idea of Foucault for whom ethics meant *"the self's relationship to itself."* It means that ethics is a kind of self-formative activity. (Davidson, 1989) The idea has an interesting relation with the view of the Russian Colonel Varnavskikh who says that *"military pedagogical practice must more actively shape this facet [the self-regulation sphere] of the individuality of the military professional."* (Varnavskikh, 2002) MPED should be more the pedagogy of creative self-development and self-regulation.

Individuality and thinking

In the contemporary operational thinking there are views that take individuality and the human dimension with a new intensity to the scene. For instance, individuality and the human dimension with ethics are highlighted as follows:

> *"The core of the joint force of 2020 will continue to be ...composed of individuals of exceptional dedication and ability. Their quality will matter as never before ... The Nation will continue to depend on talented individuals of outstanding character, committed to an ethic of selfless service."* (Joint Vision 2020)

It is the idea of RMA (Revolution in Military Affairs) that profound changes, among others, in training, leadership, and warrior skills are taking place. In recent times the discussions about RMA have turned from the plain technological viewpoint towards cultural and psychological ones, perhaps even to ethical ones. US Secretary of Defense, Donald Rumsfeld has described what he sees as fundamental RMA components: *new ways of thinking and a culture of creativity and intelligent risk taking.* (Rumsfeld, 2002)

Network Centric Warfare (NCW) is a stage in RMA. It is said that the source of the transformational combat power enabled by NCW concepts can only be understood by focusing on the relationships in warfare that take place simultaneously in the physical, the information, and the cognitive domains. The cognitive domain is the domain of the mind of the war fighter. The immaterial elements of morale, unit cohesion, level of experience, and situational awareness are elements of this domain.

One of the most important documents painting the picture of future battles, the US Army white paper *"Concepts for the Objective Force,"* says that *"soldiers are the centerpiece of our formations."* It also says that war is *"mentally complex, physically demanding, and an intensely emotional experience."* That is why *"the human dimension of warfare will always remain pre-eminent."* The term *"human dimension"* is used quite often today. Might it be that the climax of the technocentric view is over? Altogether, MPED should always be interested in the human dimension with a critical mind.

But as Roger Spiller wrote some years ago, we do not have a comprehensive theory of war on the human level. (Spiller, 2000) Scholars from many different areas have made their contributions, but still there is no theoretically coherent view of the human face of warfare. The dominating approach Spiller calls *"a materialist, instrumental conception of war."* Spiller also calls it a technocentric view. He makes a distinction between the material and human conceptions of war, and he reminds us that while the near-bloodless war is an appealing vision, the realities of modern war tell us that war is still a fundamentally human act.

From the human dimension point of view, the US Department of Defense report *"Training For Future Conflicts"* (June 2003) is interesting. The report underscores that military proficiency is as dependent on the warriors who operate weapon systems as it is on the weapon system technology. The expert team who wrote the report is quite critical about the technical view that leaves the human consequences of the military transformation unexplored. It is said in the report that transforming military training to support the warfare transformation will be a difficult undertaking. One of the most demanding things is that military personnel will have to thrive in a world and in conflicts that place more cognitive demands upon them than ever before. In other words, everybody must think on the modern battlefield.

Ethics, Self-awareness, and Agency

Especially interesting is the concept of *"self-awareness"* that is introduced into the requirements for soldiers.[21] Unfortunately, the concept is left undefined in that document, but as a philosophical and psychological term self-awareness refers to consciousness and higher-order cognitive functions like reflection and meta-cognition.

Self-awareness appears to be a central constituent in ethics. Self-awareness is the awareness of being aware; it is an awareness of one's own mental states such as perceptions, sensations, attitudes, intentions, and emotions. (Morin, 2003) To be responsible for one's own actions is impossible without self-awareness.

Alongside with individuality and ethics, a third contemporary theme is the role of teams and team building. (Brown, 2003) It is determined that combat is "based on *small unit excellence* and *well-trained soldiers* imbued with the

[21] U.S. Army (2001): Concepts for the Objective Force.

warrior ethos."[22] There are good reasons to say that we cannot have ideal teams without self-aware individual soldiers as ethical subjects.

For MPED it is fruitful to understand the relationship of self-awareness and ethics as a kind of daily work of *"taking care of oneself,"* as self-improvement, as self-mastery in the Foucaultian sense. As such it has much to do with self-discipline. And self-awareness has much to do with agency. More specifically, self-awareness and ethics are the key constituents of agency. As such, agency is the deepest meaning of *"action competence"* as the main notion of MPED. The philosopher Charles Taylor says that agency is grounded in our capacity to make *"strong evaluations,"* not only *"simple weighing of alternatives."* (Taylor, 1985: 18)[23] Immanuel Kant referred to rational agency as *Mündigkeit*, the capacity to think for oneself. Today it is not enough to say that soldiers have to think; they have to be able to make strong evaluations and they must have the capacity to think for themselves.

Ethics as Self-improvement

The term *"cognitive"* refers to psychology and to intelligence. But ethics means, first of all, critical and responsible decision-making: How should I (we) act? Ethical decisions are demanded every time we have to act in social contexts where we cannot exactly *"know"* how to act, and what kind of consequences our action will have. Ethical decision making always goes beyond *"knowledge"* that can be known and taught in advance. There can not be any routines for ethical decision-making.

More than knowledge and cognitive skills, practical wisdom is needed in ethical decision-making. Practical wisdom refers to the character of a person. This is the idea of the so called virtue ethics. We are able to be ethical only if we have a virtuous character. According to this view, the basic and everlasting task of MPED is to be education and self-education of character and practical wisdom. Practical wisdom and character are not individualistic qualities; they belong to *"Sittlichkeit,"* the substantial ethos. The basic idea is the facticity of our *"being-in-the-world."* That means that we always already are situated in a certain historical, cultural, and social reality; and that is why the only possible way to start individual self-improvement is habituation, growth of individuals into the ethos of the community.

---

22 U.S. Army (2003): The Army Future Force: Decisive 21st Century Landpower. US Army Training and Doctrine Command, August 2003.

23 *"In weak evaluation, for something to be judged good it is sufficient that it be desired, whereas in strong evaluation there is also a use of 'good' or some other evaluative term for which being desired is not sufficient; indeed some desires or desired consummations can be judged as bad, base, ignoble, trivial, superficial, unworthy and so on."*

For Michael Foucault (1986) ethics was a study of the self's relationship to itself. (Davidson, 1989) In addition to ethics, morals consist, according to Foucault, of (1) people's actual behavior; (2) their morally relevant actions, and (3) the moral code which is imposed on them. A moral code consists of the rules that determine which actions are forbidden, permitted, or required in the community. It was Foucault's interest to shift the emphasis to ethics, to the problem *"how the individual is supposed to constitute himself as a moral subject of his own actions."*

In the ethics of Foucault, there are four main aspects: ethical substance, mode of subjection, self-forming activity, and the *telos*.

(1) Ethical substance is that part of us which is taken to be the relevant domain for ethical judgment. It is like the material that is going to be worked over by ethics: acts, desires, etc. (2) The mode of subjection is the way in which people are invited or incited to recognize their moral obligations, for instance the divine law or the Kantian universal rule. (3) The third aspect called self-formative activity, Foucault describes *"asceticism"* in a broad sense. With the Greek word *askesis* Foucault means *"training of oneself by oneself."* Especially *askesis* is needed for learning *"the art of living"* (*techne tou biou*). The other Greek word (4) *telos* means the aim: *"Which is the kind of being to which we aspire when we behave in a moral way?"*

The model of ethics described above has a strong individual emphasis. But it has to be noticed that ethics is one of the four parts of morals. The other three parts, actual behavior, morally relevant actions, and moral code, are determined by society, culture, religion etc. Ethics is an individual's personal working on these elements. It is a process of study, learning, and growth.

Basically ethics means critical self-awareness of the individual. That is the only way the culture of a community and of an organization can become self-aware. Self-awareness is the key for development both on the individual and organizational as well as community levels. Using Foucault's ideas, the interactions between culture, ethos, and ethics can be collected in a model presented in picture two[24].

[24] Based on the model presented by Arnold I. Davidson (Davidson, 1989:229).

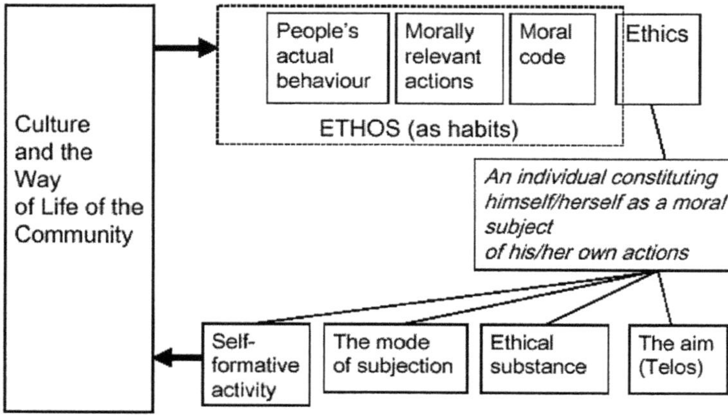

Picture 2: Relations between culture, ethos, and ethics as self-constitution.

But where, in concrete terms, is ethics needed? One example is offered by Zelenkov, when he analyzes *"the foreign countries,"* especially NATO members. His list of some negative aspects of morale includes over emphasis on material incentives, disrespect for the enemy and a trend to overestimate one's strength, racial prejudice, extreme forms of individualism, careerism, alienation and tension in interpersonal relations, alcohol and drug abuse, sexual harassment of female officers, and sexual perversion. (Zelenkov, 2001)

Zelenkov does not speak of ethics; he describes some features of morale. With his large historical material from the cruelties of the 20th century, the British historian and philosopher Jonathan Glover goes into the core of ethics: *"If we are serious about wanting to grow away from our barbaric past, this extra dimension has to be made central to ethics."* (Glover, 2001: 42) In the battle environment, especially in close combat, the moral resources of soldiers are heavily contested and they may be eroded. It is a real challenge for MPED to inquire into the problem of how to construct, support, and defend the moral identity and the moral resources of soldiers.

Ethical Viewpoint in Military Pedagogy

The field of MPED consists of the learning and growth of military people. MPED is interested in the philosophies, theories, doctrines, and practical methods of military education and training. It is not the very task of MPED to inquire into the ethics of warfare. That is the task of military ethics (ME). In principle, there is a distinction between ME and ethics in MPED, but in practice, the difference is not so clear.

There is no one universally accepted definition of ethics. Two broadly different approaches into ethics can be seen. The first approach sees ethics as *"moral philosophy,"* an objective study of moral norms. According to the other approach ethics belong tightly to practical human life, especially to decision-making and action. That is the path into ethics in MPED that I have tried to follow. According to the practical view, ethics it is not a science but a part of our life as an everlasting battle between good and evil. Ethics refers to our conscious way to conduct this battle and to construct ourselves as persons, as moral subjects of our action.

Ethics in MPED is a kind of professional ethics. From its beginning in medicine, the newly-discovered discipline of professional ethics soon expanded to law, business, and a variety of other professions, including the military. (Kennedy, 2002; Banks, 2004; Kemp, 1999) For instance, it is argued that there are four main categories of army expertise: military-technical, human development, political-social, and ethical-moral. According to the same author, the profession's educational priority is to inculcate virtues that support individual self-awareness and adaptation. (Lacquement, 2003)

In relation to ethics, there can be seen three clusters in MPED. The first one consists of the contents and methods of teaching military ethics. The second cluster consists of the ethics of MPED in itself: What does it mean to give ethically sound military training?

The third cluster, combining those two others together, should be considered the most important. It is the cluster of the very ethical challenge: How to further humanity in future conflicts by educating military people? Here we could use the terms of *"moral resources"* and *"moral identity"* introduced by Jonathan Glover. He severely reminds us about the erosion of moral identity in the environments of war. It seems impossible to see any other way to moral identity and moral resources than self-education and self-improvement. We cannot be *"trained"* to be ethical persons with a healthy moral identity and with strong moral resources.

A couple of years ago a researcher on military strategy wrote that the information age is changing the pool of manpower from which the military draws its future leaders: The new generation *"may lack discipline, motivation, and morals, but it is probably the most technologically literate cohort in the world."* (Mahnken, 2000)

Should we agree with the opinion that for military education and training it is most important to develop technological expertise while discipline, motivation, and morals are not so important? If so, we take the techno-centric point of view. On the other hand, the above mentioned paper on the Objective Force says that soldiers must have a warrior ethos, and leaders must feel *"a moral obligation"* to train soldiers to become *"tough, disciplined, and motivated."* So, concepts like discipline, motivation, and morals form an integral part of the new operational discourse.

For MPED it is important to understand ethics as lively thinking and acting in concrete situations. That was the idea of *phronesis* or practical wisdom in the ethics of Aristotle. Werner Freistetter says that *"ethical thinking transcends mere legality or even the concept of sheer 'external' obedience to norms and values, and points to an 'inner sphere' of human individuality."* And he goes on saying that *"ethical thinking cannot avoid the problem of how we are qualifying ourselves in deciding between right or wrong, good or bad."* (Freistetter, 2003) In other words, ethics cannot be but self-education and self-improvement.

Today it seems to be quite common a way to use the word *"morale"* to refer to those values and norms of the community which conduct our behavior. Always there is some moral behind our behavior, although we are not aware of it. We can talk about moral education when we are training people to behave according to some values and norms. It could be ethical education as well, but on the other hand, we can afford moral education without ethics. In that case we are training, or drilling people to behave according to some customs and habits so that they ate not aware of them. That is called indoctrination. It becomes ethical only with understanding and with freedom of choice.

We can, for instance, instruct soldiers how to behave at a checkpoint. If we only teach the routine modes of behavior, our instruction does not include ethics. But if we aim at developing their mental and social capacities to act as responsible persons in unanticipated situations and contexts, we are dealing very much with ethics. Of course, all the soldiers have to be trained to be qualified in many procedures and modes of performance. But they have to be educated as well. That means self-awareness and self-development. When trying to understand ethics in MPED, it seems reasonable in the complex field of various theories to concentrate on the *"agent-focused ethics."* (Banks, 2004) Both of them are modes of virtue ethics.

References

Banks, S. (2004): Ethics, Accountability and the Social Professions. Houndmills, Palgrave Macmillan.

Bellany, I. (2003): Men at War: The Sources of Morale. RUSI Journal, Feb 2003.

Brown, F.J. (2003): Three Revolutions: From Training to Learning and Team Building. Military Review, July-August 2003.

Casey, C. (1995): Work, Self and Society. London, Routledge.

Chiu, C-Y.; Hong, Y-Y. & Dweck, C. (1994): Toward an integrative model of personality and intelligence: a general framework and some preliminary steps. In: Sternberg, R.; Ruzgis, P. (Eds.): Personality and Intelligence. Cambridge, Cambridge University Press.

Davidson, A.I. (1989): Archaeology, Genealogy, Ethics. In: Hoy, D.C. (Ed.): Foucault: A Critical Reader. Oxford, Basil Blackwell.

Evans, M.; Ryan, A. (2000): The Human Face of Warfare. Killing, Fear & Chaos in Battle. St Leonards, Allen & Unwin.

Foucault, M. (1986): On the Genealogy of Ethics. In: Rabinow, P. (Ed.): The Foucault Reader. London, Penguin Books.

Freistetter, W. (2003): Conscience and Authority - Virtues and Pitfalls of Military Obedience. In: Micewski, E. (Ed.): Civil-Military Aspects of Military Ethics. National Defense Academy, Vienna.

Friedman, G.; Friedman, M. (1998): The Future of War. New York, St. Martin's Griffin.

French, S. F. (2002): The Warrior's Code. http://www.usafa.af.mil/jscope/JSCOPE02/French 02.html

Glover, J. (2001): Humanity. A Moral History of the Twentieth Century. London, Pimlico.

Griffith, P. (2000): Fighting spirit: leadership and morale on the 'empty battlefield' of the future. In: Evans, M.; Ryan, A. (Eds.): The Human Face of Warfare. Killing, Fear & Chaos in Battle. St Leonards, Allen & Unwin.

Hacking, I. (1989): Self-Improvement. In: Hoy, D.C. (Ed.): Foucault: A Critical Reader. Oxford, Basil Blackwell.

Honoré, R.L., Cerjan, R.P. (2003): Warrior Ethos: Soul of an infantryman. TRADOC News Service. http://www-tradoc.army.mil/pao/Web_specials/warriorEthos.

Johnston, J.; Leibrecht, B.; Holder, L.; Coffey, R. & Quinkert, K. (2002): Training for Future Operations. US Army Research Institute, Special Reports 53.

Jurist, E.J. (2000): Beyond Hegel and Nietzsche. Philosophy, Culture, and Agency. London, the MIT Press.

Kemp, P. (1999): From ethics to bioethics. In: Kearney, R.; Dooley, M. (Eds.): Questioning Ethics. Contemporary Debates in Philosophy. London, Routledge.

Kennedy, R.G. (2002): Does Professional Ethics Fail the Profession of Arms? Presented to the Joint Services Conference on Professional Ethics Washington, DC, 24 January 2002.

Koehn, D. (1994): The Ground of Professional Ethics. London, Routledge.

Lacquement, R.A. (2003): Understanding professional expertise and jurisdiction. Military Review, March-April 2003.

Laitinen, A. (2003): Strong Evaluation Without Sources. On Charles Taylor's Philosophical Anthropology and Cultural Moral Realism. Jyväskylä, Jyväskylä Studies in Education, Psychology and Social Research, No 224.

Mahnken, T. (2000): War and Culture in the Information Age. Strategic Review. Winter 2000.

Morin, A. (2003): A neuro-socio-cognitive model of self-awareness with an emphasis on inner speech. http://cogprints.ecs.soton.ac.uk/archive/00002832/01/SAmodel.pdf.

Neill, D.A. (2000): Ethics and the Military Corporation. Canadian Military Journal, No 1.

Noskov, Y. (2002): Military Ethical Reform in Eastern Europe and Russia. Presentation at International Military Ethic Symposium Riga, Latvia, 5 - 10 March. http://www.accts.org/ethics/latvia/titles.htm

Proudfoot, M. (ed.) (2003): The Philosophy of Body. Oxford, Blackwell.

Rabinow, P. (ed.) (1986): The Foucault Reader. London, Penguin Books.

Rumsfeld, D. (2002): Secretary Rumsfeld Speaks on "21st Century Transformation" of U.S. Armed Forces. http://www.defenselink.mil/speeches/2002/s20020131-secdef.html.

Spiller, R. (2000): Introduction. In: Evans, M.; Ryan, A. (Ed.): The Human Face of Warfare. St Leonards, Allen & Unwin.

Taylor, C. (1985): Human Agency and Language: Philosophical Papers. Vol. 1, Cambridge, Cambridge University Press.

Taylor, C. (1995): Philosophical Arguments. Cambridge, Harvard University Press.

Toiskallio, J. (2003): Ethics, Military Pedagogy, and Action Competence. In Micewski, E. (ed.): Civil-Military Aspects of Military Ethics. National Defense Academy, Vienna.

US Army (2001): Concepts for the Objective Force.

US Army (2003): The Army Future Force: Decisive 21st Century Landpower. US Army Training and Doctrine Command, August 2003.

Varnavskikh, E.A. (2002): Molding the Individuality of the Military Professional: A Theoretical Aspect. Military Thought.

Zelenkov, M. Yu. (2001): Morale Training in Foreign Armies. Military Thought, Jan, 2001.

Peter A. MATTSSON, Swedish National Defence College, Stockholm

NETWORK CENTRIC WARFARE - AN ANALYSIS OF SOME PEDAGOGICAL IMPLICATIONS

Abstract

The aim of this paper is to analyze and to discuss the four basic hypotheses of the Network Centric Warfare of US Defense Forces in a military pedagogical perspective[25]:

1) A robustly networked force improves information sharing.
2) Information sharing enhances the quality of information and shared situational awareness.
3) Shared situational awareness enables collaboration and self-synchronization, and enhances sustainability and speed of command.
4) These [hypotheses], in turn, dramatically increase mission effectiveness.

Finally, there is a discussion in progress of the ambition and size of the Swedish version of NCW, which is called Network Based Defense (NBF in Swedish), as well as of some of the pedagogical implications for the Armed Forces in Sweden.

> *"We must build forces that draw upon the revolutionary advances in the technology of war ... one that relies more heavily on stealth, precision weaponry, and INFORMATION TECHNOLOGIES."*
>
> George W. Bush, Commander in Chief (The Network Centric Warfare Report, 2001: 1)

Network Centric Warfare - Strategic Implications

The Swedish Parliament has decided that Sweden shall create a network-centric defense. The reasons for taking this course are the new opportunities of the information age and the debate that has taken place, especially in the United States, about a revolution in military affairs (RMA), partly based on communication technology.[26] Therefore, the Swedish Armed Forces shall

---

[25] A military pedagogical perspective is to be understood as a scientific critical perspective based on general epistemological and pedagogical theories and empirical results.

[26] The Swedish term 'nätverksbaserat försvar' (NBF = network-based defence) originates mainly from the American term *Network Centric Warfare* (NCW) but it is focused on defense not warfare. The same applies to the Swedish term 'military revolution' which in English is called *Revolution in Military Affairs* (RMA). The information age is an

develop capabilities relating to information and also the knowledge of the intelligence, command and decision support, as well as engagement and impact. (Bill 2001/02: 10; 127-129)

The United States Congress has decided to introduce network-centric warfare. The purpose of American network-centric warfare is to improve their vital processes from information-analysis-interpretation-decision-action and to decrease defense costs. When all this is taken together, the total effectiveness of the armed forces is expected to increase. The idea comes from the realm of global business, which has used communication technology to increase the competitiveness of companies. (Money, 2001)

This way of thinking is called revolutionary because the assumption is that it will change warfare dramatically. Even the military view of knowledge is more revolutionary than the long dominant experience-based view. All of this raises questions about views on knowledge and about learning and change, individually and collectively.

This article will first provide a brief background on the American concept of network-centric warfare (NCW) as well as the Swedish concept of network-centric defense, in order to conduct a pedagogical examination of the fundamental American hypotheses on NCW.

Joint Vision 2010/2020 - An Offensive Doctrine

In the American doctrine, *Joint Vision 2010/2020*, there is an ambition to create joint armed forces which, put concisely, shall be persuasive in peace and decisive in war. The Army, Navy, Air Force, and Marine Corps shall coordinate to achieve full spectrum dominance by operating jointly and by sharing resources. The vision describes how the American armed forces will become faster and achieve more efficient and precise combat.

This transformation and improved efficiency shall be enabled by utilizing an expected potential of communication technology that is expected to increase military capabilities. According to the vision, all this change shall lead to dominant maneuver, precision engagement, focused logistics and full dimensional protection. Achieving these aims requires a complete coordination of the armed forces: intellectually, operationally, organizationally, doctrinally, and technically.

Together with intellectual and technical innovations, the communication technology that is currently available shall enable this transformation from

---

overarching term to describe the modern industrial society's transition to an information society (See e.g. Castells, M. (1996): *The Information Age. Economy, Society and Culture*).

platform-centric to network-centric armed forces. (Joint Chiefs of Staffs 2000: 1-4).

Network-Centric Warfare (NCW) [27]

The American discussion about modern defense has resulted in decisions to implement comprehensive changes in the coming decades. More and more commentators believe that the United States will have to utilize modern, civilian communication technology if it is to maintain its military dominance. In the United States it has principally been the Air Force, Navy, and the Marine Corps that have been behind this development.[28]

The American thinking on network-centric warfare has been developed based on four premises (FY01, SEC.934i):

- A robustly networked force improves information sharing.
- Information sharing enhances the quality of information and shared situational awareness.
- Shared situational awareness enables collaboration and self-synchronization, and enhances sustainability and speed of command.
- These [premises], in turn, dramatically increase mission effectiveness.

The reason for these dramatic changes in direction and these transformations of the armed forces in both the United States and Sweden is said to be the transition of the industrial society to the knowledge-based and information society. The information age is the transformation of the agrarian and the industrial societies' economic focus on physical resources to a new focus on information and knowledge. It was in the beginning of the 1990s when the discussion started about a new form of competitive companies and businesses that utilized communication technology.

---

[27] The American term 'Network Centric Warfare' has been defined in Sweden as Network Centric Defence (Swedish term: Nätverksbaserat försvar or NBF).

[28] From 1997 to 1999 the American Navy and Marine Corps published several reports at conferences: the most notable was probably *Network Centric warfare - Its Origin and Future* by Admiral Arthur K. Cebrowski, U.S. Navy and by the military researcher and adviser John J. Garstka from the U.S. Naval Institute 1998. (see http://www.usni.org/Proceedings/Articles98/PROcebrowski.htm). Another important participant in these discussions is *The Command and Control Research Program* (CCRP) which is part of the U.S. Department of Defense (DoD) and which aims at creating contacts between the civilian and the military use of management/command methods and communication technology (http://www.dodccrp.org/). A decisive step towards network-centric warfare in the U.S. is *the Network Centric Warfare Report* which was submitted to the U.S. Congress in July 2001 (http://www.dodccrp.org/NCW/NCW_report/flash.htm).

This new thinking on virtual companies whose principal assets are ideas and innovations was based on market knowledge and knowledge of which companies could develop and supply these products and services. These entrepreneurs could be seen as a form of knowledge brokers who used the connection of rapid market and customer analyses to develop new, competitive products and services. (Davidow & Malone, 1992)

Consultants and researchers began to issue warnings that traditionally strong American companies could be wiped out by these new virtual companies that made clever use of modern communication technology to dramatically increase their competitiveness. The American experiences of technological ventures during the 1980s had given varying results with numerous expensive failures.

Consequently, there was no lack of skepticism towards this revolutionary new communication technology. Especially in Europe, the values behind this new revolution were questioned. The results in the United States showed dramatic changes, especially regarding the reduction in staff and the discontinuation of various products and services, which were purchased from other companies. However, the American economy was much stronger and more dynamic than the stagnated European economic development during the 1990s. One of the most important competitive factors was the implementation of communication and information technology. (Coulson-Thomas, 1994)

But the important lessons of what came to be called *re-engineering* and *business process re-engineering* were that technology must always be introduced parallel with changes to the company's business concept, corporate strategy, and its most important processes. A holistic perspective and a shared loyalty to the new thinking were important. This pointed to difficulties in managing phenomena as old experiences and resisting people that did not fit into the new business. (McHugh, P.; Merli, G. & Wheeler III, W.A., 1995)

An Examination of the NCW Hypotheses

When reading the American descriptions of NCW, it becomes apparent that there is a lack of references and applied research. So, it is a challenge to review as well as to analyze and to discuss the basic hypotheses on which the entire concept is based: *"A robustly networked force improves information sharing."*

This hypothesis can be considered to be valid if the networked resources mutually provide information to each other. Technically the network can be created and developed to bind together systems that were previously incompatible. However, this hypothesis assumes that there is a desire to share and thereby also an added value in sharing.

The pedagogical challenge lies in creating the conditions whereby such a mutual desire is developed with an aim to optimizing the sharing of information. Such a statement indicates that the technical system in itself is insufficient to forward all the information if people do not cooperate and are not motivated to create an informational added value. The implementation of automatic sensors and system testing systems can to a degree create a certain added value in the formation of joint information.

Experiences from distributed project management systems and overarching management systems in the business world indicate that as long as any added value cannot be spread within the entire organization, reporting and information distribution declines and sometimes ceases. It is a pedagogical opportunity to shed light on this behavior and to develop organizational feedback systems that reinforce the individual's own desire and ability to provide information and knowledge in a network.

There are concerns that the military solution to this undesirable behavior is a system of force and sanctions. Technical systems give traceability, which can be utilized to create obedience and conformity in networks. Therefore, the openness of the technical systems can have the opposite effect, so that it instead oppresses initiative.

The success of introducing shared information systems in networks will probably be decided by the ability of the higher military command to listen to the needs of the lower levels. A successful approach like this requires several positive experiences and fruitful experiments and trials. This can provide the success necessary and can create momentum for a continued favorable development.

Information Sharing – Quality of Information and Shared Situational Awareness

This premise appears more uncertain than the first hypothesis. Partly because its validity is based on the validity of the first hypothesis and partly because it seems to assume *a priori* that information sharing (presumptively a large amount of information) will increase the quality of the total information and provide a joint situational awareness.

The sharing of a large amount of information could mean that the quality of the information increases but this assumes that there is a capability or a capacity within the imagined network to refine information into potential knowledge, which in turn can lead to decisions. Here it is interesting to ask, where does the quality of the information increase and for whom can this be decided?

Joint situational awareness is a logical impossibility. Such intersubjectivity cannot be fully achieved. However, systematic work with doctrines and advanced military abstract thinking can provide important joint frames of reference that can be used in experiments and trials. These experiments and trials should develop new empirical knowledge on the part of military officers and their associates.

During phases of problem-solving, various methods are used, of which the first is a pure form of assimilation, i.e. each individual tries to respond to the new situation's stimuli and information with the accumulated knowledge and experience that the individual possesses. Assimilation is an economical way of achieving a balance between the task andrequirement and one's own capability.

The other problem-solving method is to use the ability of other individuals to assimilate and in this way to solve the task quickly and efficiently. The third method is a form of analogical reasoning that uses frames of reference and the ability to assimilate but which can lead to one's own thought structure changing and expanding. This is a milder form of accommodation that involves a qualitative change of the thought structure.

Last, it is experimentation that can be based on frames of reference, in which case it becomes focused and goal-oriented. It can also be random and unsystematic. This can occur individually in the beginning but it often leads to co-operation between several individuals.

Joint situational awareness does not land on a *tabula rasa*. The military officers whose duty is to man the network defense system have a particular socio-cultural background and hopefully a cultivation of the military system. During his career a military officer develops a multi-facetted identity that stretches from the primary group of the basic officer's training to being an expert and incorporated in sophisticated military, technical, or organizational systems. The officers that participate in the NCW defense must manage their own and other's range of experience and knowledge.

Put briefly: you reap what you sow; what you have sown during the golden age of the distinct anti-invasion defense period, you will reap in the uncertain future of network-centric warfare. It is truly a pedagogical challenge to create – as far as possible, from the highest abstract military thinking to the instinctive motor behavior – an intersubjectivity which gives improved quality in shared information and situational awareness. Actions and activities must then be connected to this, but I will return to this under the last hypothesis.

Shared Situational Awareness – Self-synchronization, Enhanced Sustainability, and Speed of Command

This hypothesis is based on, and therefore presumes, the validity or at least the plausibility of the two previous hypotheses. It would be reasonable to assume that joint situational awareness (to the extent this is possible) can strengthen cooperation. This presumes that everyone who participates in the activity wants to take part in a joint situational awareness and wants to cooperate according to the instructions or requirements that are expressed in the network. It is possible that the openness (transparency) of the network will clearly reveal differences in competence within the network and that this might limit the desire to cooperate. Or could it be the other way around?

This raises new questions that originate in the requirement to cooperate. Which type of cooperation is intended and what does this cooperation involve as far as the competence of the network's members? Cooperation should cross the boundaries of services and functions, according to the American ideas. Fragmentation and segmentation must be counteracted and joint ways of thinking and acting must be developed. How should these be developed: in independent network units or in controlled activities?

> *"Organizations must continually reinvent and update their common knowledge. This requires them to engage repeatedly in two kinds of knowledge activities. First, they have to find effective ways to translate their ongoing experience into knowledge - create common knowledge. Second, they have to transfer that knowledge across time and space - leverage common knowledge."* (Dixon, 2000: 17)

The American pedagogue Nancy Dixon has, in her research on common knowledge in multinational companies, studied these companies' ability to handle knowledge formation internally and externally. The process of the organizational knowledge work begins when current knowledge is identified as well as the knowledge that is missing. The next step is to distribute current knowledge and to develop new knowledge by internal and external resources. The third phase is most difficult when knowledge has to be transferred from one individual to others or from one research group to others. Employees that have not been active in the creation of knowledge have always difficulties in assimilating new knowledge. (Dixon, 2000)

According to Fichtner, cooperation can be expressed in three levels: for instance coordination, joint development and reflective communication, which coincide with the three-act levels consisting of operations, actions and activities. Individually constructed knowledge does not acquire value until it is tested collectively. However, individuals must not become standardized by collective thought models because collective development of knowledge is based on shared mental models which are essentially dependent on new and boundary-spanning knowledge from its individual members and the surroundings in which they are active. Engeström argues that individual operations and acts must change into collectively boundary-spanning activities in order to be able to expand collective and thus organizational knowledge. (Engeström, 1987: 333-334)

Perhaps the collective formation of knowledge and the open, knowledge-expanding organization are what are meant by the increased level of cooperation. The question would then have to be whether this is a military aim or whether this (terrible) thought has become apparent. The open network organization controlled by a technical system would result in an obvious change and hard consequences for a traditional military system and a hierarchical organization.

In relation to self-synchronization, there are a number of delimitations missing from the descriptions. What is it that is to be self-synchronized? Is it subsystems in larger systems, i.e. a form of serial self-synchronization because a system must have a purpose and a given function? Synchronization can mean concurrence, cooperation or coincidence in time and space. This indicates that synchronization is intentional. So who determines the boundaries of a self-synchronizing system and its synchronization with other self-synchronizing systems that are linked to an overarching system?

This seems undeniably like a form of management tool or a principle of management science with its roots in systems theory and cybernetics. Self-regulating systems are controlled by the will and knowledge of the system designer; others in the system are subordinated to that. Sometimes a self-regulating system is considered to be an intelligent system with sensors and regulators that can get it to start processes, for instance a heating unit. Heating units can be given additional functions, which makes them multi-functional but they are still limited by the input and the function that is assigned. Beyond these input values, it cannot act intelligently. In fact, we can suspect that such a system can be misled, for instance by IR sensors and IR alarms.

There may be ideas about self-synchronized human systems or functions. The same questions can be asked about these. As well as other questions about what will and morale can achieve in self-regulating systems of people.

Increase Mission Effectiveness

This is the most uncertain hypothesis of them all and it presumes that the three first ones are valid or at least probable. From the analysis of the three previous hypotheses, the balance between the technical and the non-technical parts of network-centric warfare has been most important. In the American descriptions of NCW, the physical domain seems to be the starting point, then this changes into the information domain and ends in the cognitive domain.

The physical entities in the form of network and weapon platforms seem to be most important in NCW. However, attention is paid to the general factors that control the execution of intelligence and combat systems. This raises questions about important processes such as command, education, and training which affect the soft factors like morale, endurance, and mental strength.

In the end, the track stops in the individual brain or the individual cognitive domain. A person in possession of a shared information and situational awareness who is to move into some form of self-synchronization, which, taken together, are to give dramatically higher efficiency in military and peacekeeping operations.

Dramatic changes can be achieved in the shortening of important key figures or lead times such as the time required to discover and engage targets or early growth in strength. This has been achieved through previous experiences and routines being challenged and partially abandoned for a revolutionary thinking instead of an evolutionary one. The shortening of lead times by more than 80% has been achieved. In combat and battle efficiency have increased by several hundred percent, according to certain publicized reports.

Finally, you must be humble when you try to understand and realize what NCW is all about. After reading the official documents and books you ask yourself what kind of mistakes and overwhelming successes have not been reported. What is hidden from the researcher's eyes? One thought is the thought of deception. Could these be a part of a greater form of information and psychological warfare directly and indirectly?

Or is this romantic feeling of fascination correct that we are today looking into the real military revolution of thought, knowledge, and warfare? I really don't know. Whatever it is or could be - it is interesting!

Some Pedagogical Implications

> *"The mission of 2020 will demand Service members who can create and then take advantage of intellectual and technological innovation. Individuals will be challenged by significant responsibilities at tactical levels in the organization and must be capable of making decisions with both operational and strategic implications. Our vision of full spectrum dominance and the transformation of operational capabilities has significant implications for the training and education of our people."* (Chiefs of Staff, 2000: 19)

Some of the premises and discussions of NCW is exciting and breathtaking. There is a strong faith in the American capacity of economy, technology, science, and people to manage to develop and carry out the vision of the network-centric warfare. It is not only about technology as can be seen in the NCW guidelines that express a dynamic co-evolution of doctrine, organization, technology, and people. One important pedagogical conclusion is the needs of individual and collective training in a new form of environment of jointness and communication technology.

The Swedish ambition of the NBF (Network Based Defense) is much lower than the American NCW and it is coordinated to the new Swedish doctrines and security policy. In the NBF-development project the main work has been divided into groups of technology, methods, organization, and people. The new Swedish strategic doctrine was released in June 2002 and it will be followed by the operational doctrine and the doctrines of the services in 2004.

There is quite a high pace of change within the Swedish Defense Forces, which implies a need for new pedagogical interventions. Whatever happens, there is a significant change in the way of conducting warfare and peacekeeping operations. New technology combined with a more professional and knowledgeable organization should increase the operational impact of these NBF-forces if they are appropriately supported by training and education. Pedagogically, one can wonder whether a shift from an evolutionary, experience-based view of knowledge to a revolutionary, theory-based view of knowledge will change the key figures of learning and knowledge formation within the Swedish Armed Forces.

Finally, I would like to address some pedagogical issues to be further discussed:

- How will the increased number of simulations, theoretical calculations and experiments affect the collective amount of greater knowledge within the Armed Forces (e.g. the organizational knowledge)?
- How will the senior officers be able to meet the requirement of the specific high knowledge to be able to handle, control, and command sophisticated technological systems?
- Does the great majority of officers lack a basic academic degree - a substantial organizational risk to be successful in the implementation of the NBF (which is not the case in many comparable countries)?
- Would the harder NBF-requirement result in a knowledge and competence crisis within the Swedish Armed Forces?
- How will a military system (such as the Swedish one) with such a wide range of knowledge and skills be able to manage such a radical organizational transformation that NCW and network-centric defense seems to involve?
- And what added value could pedagogical science generate to the Swedish Armed Forces during such a NBF-development?

References

Albers, D.S.; Garstka, J.J. & Stein, F.P. (1999): Network Centric Warfare. Developing and Leveraging Information Superiority. Vienna, Va., CCRP Publications Distribution Center.

Barnett, T.P.M. (1999): The Seven Deadly Sins of Network-Centric Warfare. http://205.67.218.5/dsd/7deadl~1.htm, 2002-07-14.

Coulson-Thomas, C.J. (ed.) (1994). Business Process Re-engineering: myth & reality. London: Kogan Page.

Davidov, W.H. & Malone, M.S. (1992): The Virtual Corporation. Structuring and Revitalizing the Corporation for the 21st Century. New York, NY. Harper Business.

Dixon, N.M. (2000): Common Knowledge. How Companies Thrive by Sharing What They Know. Boston, MA. Harvard Business School Press.

Engeström, Y. (1987): Learning Expanding Knowledge. An Activity Theoretical Approach to Development Research. Helsinki, Orienta Konsultit Oy.

Hoffman, F.G. & Horne, G. (eds.) (1998). Maneuver Warfare Science 1998. Quantico, VA., Marine Corps Combat Development Command.

Horne, G. & Leonardi, M. (eds.) (2001). Maneuver Warfare Science 2001. Quantico, VA., Marine Corps Combat Development Command.

McHugh, P.; Merli, G. & Wheeler, W.A.III. (1995): Beyond Business Process Reengineering. Towards the Holonic Enterprise. Chichester, John Wiley & Sons.

Money, A.L. (2001): Report on Network Centric Warfare. Sense of the Report. FY01, Public Law 106-398, Section 934, Washington, D.C., U.S. Department of Defense.

Reed, E.S.; Turiel, E. & Brown, T. (eds.) (1996): Values and Knowledge. Mahwah, NJ, Lawrence Erlbaum Associates Publishers.

Stein, F.P. (1998): Observations on the Emergence of Network Centric Warfare. Vienna, VA., Evidence Based Research, Inc. http://www.dodccrp.org/ , 2002-07-17.

Wheatley, M.J. (1992): Leadership and the New Science. Learning about Organization from an Orderly Universe. San Francisco, CA, Berett-Koehler Publishers.

Rainer P. BORN, University of Linz, Linz

Creating Sensibility in Applied Ethics

Once again we are learning that anything truly complex cannot be broken down into its parts without losing the integrity of the whole
(Verna Allee: The Future of Knowledge, 102)

Setting the Scene: Rules and Sensibility

Usually we are told: ethics and success (e.g. in business) are incompatible. Or: In business one cannot afford ethics - ethics is luxury. Pity is dispensable. But that surely depends on the intuitions concerning one's life, *"on what we aim at and how we [want to] live."* (Cf Julius Moravcsik)

In *"Fugitive Pieces"* Anne Michaels describes the situation of two lovers: *"It is late, almost afternoon, when she says, though I may have dreamed it, though it's just something Michaela might ask: Are you hungry? No. ... Then perhaps we should eat so that hunger won't seem, even for a moment, the stronger feeling."* Perhaps this is what ethics is about - not eating but thinking ahead, empathy, sympathy, thinking of others taking care of a common future - but this is just a beginning.

Usually in considering ethics one starts with some dictionary definition(s) about ethics or about what **is** to be considered as good. Rarely one discusses a/the task(s) of ethics or why one should consider *"the good"* or even to be good. Definitions that are like cleaning up an empty room, that tell us nothing new, that tell us something that has to fit within what we already know (more or less) or seems to be compatible with an established world view. In the sequel this paper shall try to argue critically against such an attitude because being alert is essential in this world now. Look at the following picture (Fig. 1) - what could it tell us by analogy?

The point to be stressed is that it could establish some kind of "sensibility" concerning the misinterpretations of maps (as far as they are used as action-guiding). It is the deliberate reduction or simply the loss of dimensions which leads to misapplications of our (ethical) theories, models, or charts of the world. The loss of dimensions may imply the loss of solutions to problems and of possibilities to "correct" the coming about of un-intended results. Thus it is not always the case that the direct or shortest path (according to the map) to reach some aim is optimal or is the best *"solution"* to some problem.

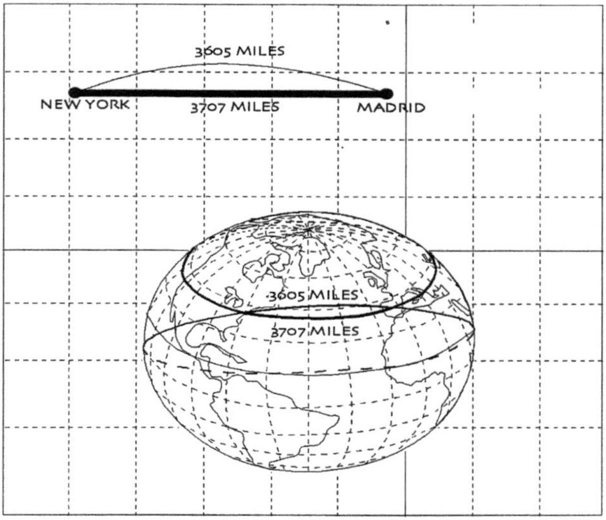

Fig. 1: Taking Care of Dimensions

In the sequel an approach will be taken up which in the meantime was already explained by J. Moravcsik (he calls it an alternative to utilitarian and Kantian ethics as classical approaches). This approach shall be extended accordingly, i.e. to fit into own model-theoretical studies. The aim is to take into account both the way in which we are affected by the actions of other people as well as the other way around when the fulfillment of our duties and the responsibilities concerning the consequences of our actions are to be taken into account.

Concerning the topic what we or others ought to do in some situation Moravcsik argues in the following way:

i) *"On the one hand, we can be harmed by actions and circumstances that work against realizing what is in* ***our interest*** *and what would help fulfill our potentialities."*

ii) *"On the other hand, our actions and attitudes can harm or diminish our contributions to the benefit of others."*

According to Moravcsik these *"dangers"* can be divided with respect to the prevailing ethical theories, into two groups:

- The first [danger] is concerned with *"what we want to do with our lives and how we find* ***means*** *to carry out such* ***plans,"***
- The second [danger] concerns *"matters of morality, to be specified in terms of an autonomous set of* ***rules."***

Furthermore I have to stress again that I am not starting with a definition of ethics. If at all, I shall provide a sort of *"implicit definition"* (in the sense of D. Hilbert & M. Schlick), depending on tacit background knowledge or intuitions of ethics. I am much more concerned with what ethics is about, what kind of help it can be what kind of role it should play and what kind of task it should fulfill. So the question is: What kind of help do we expect from codifying ethics in a set of rules (or structural regularities concerning situations in which we have to act)? Thus I should point out that ethics concerns our *"sensibility to reflectively judge our action-guiding rules and measures on the basis of both duties and responsibilities as far as they are concerning the consequences/ outcomes/values of our actions."*

Another way to create some *sensibility* may be the following situation which I call *"b(l)ending the rules:"* The traffic situation explained in the picture is an example of IN-Sensibility (ethical and otherwise). The best solution to unload the truck is not the best for the whole situation. The people unloading the truck do not take care of the other people. But the main problem is not solely the depicted traffic situation. The main problem is whether it would be possible to solve the situation with the help of a set of codified rules to act appropriately plus some sanctions (s.b. *"the paradox of the Samaritan"*) if those rules are not obeyed. Could this set of rules be made complete or do we need some way of reflective correction in order to remain flexible, adaptive and innovative with respect to future situations? A suggestion for a solution follows below.

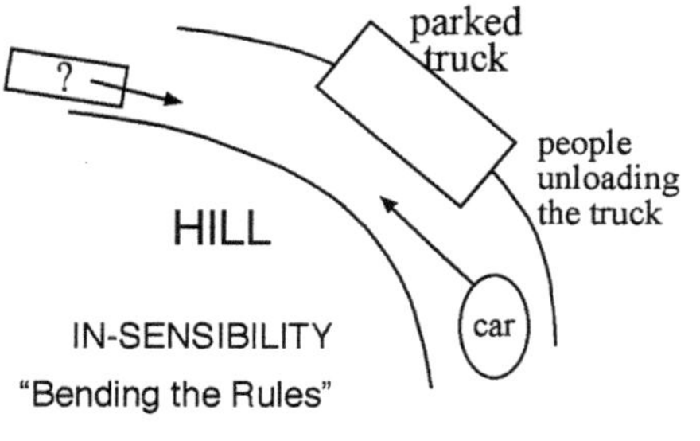

Fig. 2: B(l)ending the Rules

I shall first discuss two examples:

i) Consider a big aircraft carrier. What kind of organization does one need? Karl Weick (organizational psychology) calls it an HRO (high reliability organization) because the slightest mistake (forgetting a screw driver on the landing strip or other so called minor incidents) can have disastrous consequences. But do you think it is just rules for organizing and sanctions for not obeying rules which will settle the matter? I think it is extremely important to understand what one is doing, not just to be motivated but to see the sense in what one is doing and to see the possible consequences of ones actions. You cannot even reduce it to machines because there always may be some unexpected events or situations. So rules are not enough.

ii) Think of a child, perhaps 10 years old or so, using a pocket calculator to imitate a computation you have shown to her/him, e.g. the size and prize for a full coverage carpet of the floor of a certain room. The child forgets some decimals and the result is much too large a number. But the child has no experience, no feeling, and no interpretation of the numbers in some context. The children have not yet a sense about what they are really doing. And even if the computed result is correct it may be wise not to assume that the child has been *"computing"* the correct solution unless we are sure it has understood what it is all about so that in some critical future situation we can really rely upon the result. The same is true if we just obey *"ethical"* rules with no sensitivity concerning the consequences of their application.

Of course ethics does not mean that we should be like centipedes not knowing any more how to walk once we have been asked about the rules of moving and reflecting all the time. Actually it should be the other way around. We should be very sure and safe and *"highly reliable"* in normal situations and still have developed a feeling or sensitivity for unexpected or limiting cases where it becomes necessary to act quickly and still responsibly, creatively, innovatively, fast, and successful. That is what a good and ethically reflected behavior should be about.

Furthermore it should be stressed that rules need to be expanded by considerations concerning emotions (Cf Arno Gruen) and sensibility. This aspect is already available in Plato but was more or less narrowed down in the reception of the classical approaches, i.e. especially Kant and utilitarianism.

Thus in the sequel it shall be argued (and insisted) that on the one hand it is essential to realize that a basic ethical attitude toward our human fellows can be grasped or replaced neither by a rigid set of rules nor can it be codified or be built up by such asset because a successful use of norms and rules already presupposes some background knowledge or some responsible basic attitude in order to guide or steer our actions. Last but not least we should aim at sensitivity concerning situations [in which case for analyzing arguments I am thinking of *"situation semantics"* - Barwise/Perry/Etchemdy] that by no means needs to lead into the centipede paradox. A modern example (which also concerns the misguided term change of values) of the topic are so called heart transplants. If you take the Hippocratic Oath seriously and conclude you have to save life in any circumstances you are in a dilemma. On the other hand it is possible to save life. Thus in order to justify the medical measure one has to replace the intuitive notion of death by heart by *"brain-death,"* thus degrading our heart to a *"heartless, unfeeling mechanical pump."* Thus the original *"value"* to save life could be kept up and only the set of *"possible actions"* to reach the goal had to be changed or rather there had to be established new *"constraints"* concerning restrictions with respect to medically possible measures (e.g. euthanasia), i.e. identifying *"admissible"* measures to produce a subset (ethically justified) of *"good"* actions. Insofar as ethics concerns these constraints it has a highly empirical impact both positive as negative, if you think of certain physicians in concentration camps in the national-socialist period. As I shall argue in the X-ray example below it is not enough to learn rules and create some scientific understanding of what is going on in a body in studying medicine but one also has to develop one's ethical sensibility.

Concerning the situation between possible and admissible solutions the following graphic may be helpful:

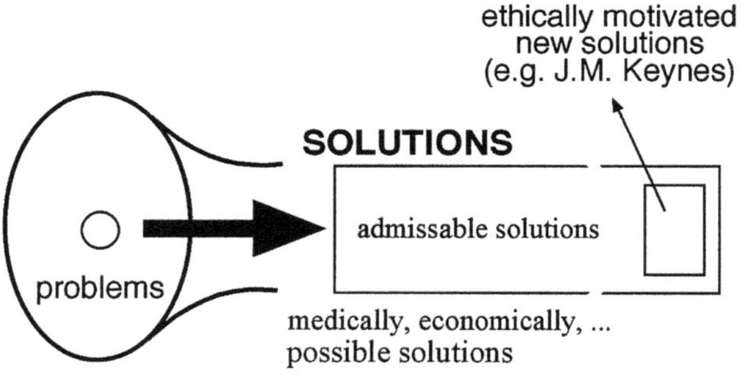

Fig. 3a: Admissible Solutions

The main problem seems to select (and reproduce) the set of admissible actions to be considered as problem solutions starting with the set of computationally, logically, or logistically possible decisions concerning actions; i.e. one wants to set up a set of rules ex post facto, a set that both reproduces and explains the admissible results. In this way so-called standard or default solutions (depending on accepted background knowledge) will become understandable by argument and will be reproducible in a controlled manner. But concerning the *"value"* of new solutions one has to establish a sort of feeling, a sensibility, an eye for the context, i.e. to know how far one should go (or would be allowed to go) in a certain case and whether one should confide in the given rules or whether one should keep in mind some measures of correction. This is where sensibility, experience, knowledge, and thinking are needed.

Basically one has (not always consciously) to evaluate ones actions in their coming about (which also has to take into account ones duties) as well with respect to their consequences (problem of responsibility) and has to justify decisions according to their consequences.

Further Considerations to Create Sensibility

Just for arguments sake suppose that in a jumbo-jet both pilots are sick and cannot fly any more. The plane is now under control by auto-pilot and a 12 year old child in charge, a child with sufficient experience in playing a flight simulator computer game and in audio contact with the tower of the next airport. The passengers are just being informed. How do they feel?

Now what has this to do with ethics? Do we think that the child can take responsibility of the plane? Is the tower able to support the child sufficiently? Do we trust the techniques?

Who actually does have responsibility concerning the plane and a secure landing? Who does take responsibility for the life of the passengers? Do you think it is the child? (What would lawyers think of it?) Could we say the autopilot or the programmer or other persons are responsible? Do we think the autopilot is a *"complete"* system and what kind of wrong expectations do we have in trusting in the program? Now there is some parallel to ethics!

Should we believe that an ethical rule system can be complete? Do we think there is no technical correction necessary, no interpretation (of instruments or data in case of the pilot), no understanding or sense making? Asking differently, do we not think that it is necessary to have experience and sensibility? Could it not be that in trying to explain the success of an experienced pilot, we have to think of him being able to make best use even of an autopilot, knowing when it is reliable when to switch off because are we suddenly in a non-standard-situation!?

Any kind of experts system (even in ethics) needs experienced application! What is wrong is to believe that there could exist a very sophisticated expert system so that any person with normal intelligence (universal commonsense) could use it and produce the same result as a good expert. Which does not mean that specialists in a certain field cannot have a blind spot with respect to ethics - if you think of certain Nazi-doctors? But the main problem is creativity and innovation with respect to future developments. Not everything is settled by blind evolution or, phrased differently, it might well be that evolution has developed reflection in humans (and ethics definitely falls within this area) as a means to surpass or overcome blind evolution. *"Reason can transcend whatever reason can formalize"* (H. Putnam) is my favorite quote in connection with

Goedels incompleteness theorem concerning these matters as far as the limits of formalization and the blind or mechanical use of rules go.

But does it not hold true of both soldiers and officers that they need to know when they find themselves in a situation of exceptions where one has to evaluate and estimate some situation and has to apply the system of rules in use with some care and sensitivity?

What seems to be essential here is that rules of action need to be formulated in a way that corresponds to common sense thinking and common sensical moral intuitions and implicitly given value systems or sets of beliefs. But this does not hold true of the justification of those rules, norms, or checklists. Mixing these two up is rather the essential problem of applied ethics. Only if we really think that our future can be completely regulated or is completely determined and grasped by our system of rules or logistics would it be meaningful to expect that there will be nothing unexpected to turn up.

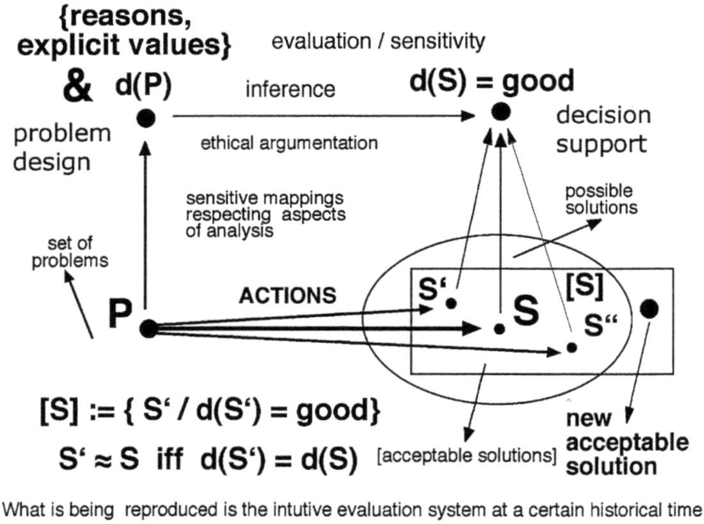

Fig. 3b: General Idea

Quite often we say that some people work well in accord with rules. But usually one does not look carefully to identify the real causes which are pre-experience and the establishment of some good background knowledge, because the latter enables us to correct mistakes of application.

The so-called alternative approach, as Moravcsik calls it, insists that a justification of ethics does not tell which kind of practical sensibility is necessary in a historical situation. In learning to use a map we have to learn both which simplifications are used (to be able to relate a map to some real landscape) as well as rules to use the map, i.e. we need to create sensibility with respect to reality. We have to learn that a map is neither a literal description of reality nor a literal device to guide our actions.

In this sense it is an important task of ethics to introduce something like a human measure concerning the use of rule systems, that is as a means of correcting overly rigid instrumental algorithmic and therefore unreflected dealing with rules. This means that so-called thinking by content use of guiding conceptions needs to be introduced into the application of rules (semantics can not be reduced to syntax). The paradox in this situation is that we have to choose our rules in such a way that in the normal field of experience they correspond to our moral feeling. Thus we could act quickly and in a reliable manner and are able to act (not JUST A CENTIPEDE who can not move anymore if being asked how he manages his legs), i.e. we need security and a good feeling about the limits of our actions. We need to become aware of the reach and the validity of experiences in standard situations. We need schooling, a lot of experience, and the creation of sensibility and responsibility and consciousness of our duties. In this way ethics in its form as moral rules (conscious or not) becomes an enormously important empirical factor in our everyday life. This leads to a second paradox: we seem to believe that sets of rules can take over the responsibility for our individual acting. But the latter would hold true only in a mean area of living and since we have to become aware of the limits of our rule-systems responsibility cannot be taken over by the latter. The practical result of too many codices in general is the fact that one wants to circumscribe rules and does not develop any moral sensibility as soon as there seems to be a personal advantage (C.f. "bending the rules").

Naturally in ethics all kinds of definitions are given which may be comprised as a sort of idea of good. In the sequel the point is to judge ethics with respect to the task it should fulfill in the context of philosophy or rather reflective orientation in the world. In principal, ethical problems turn up when we are asked to do something that we have the feeling we ought or should do. The next problem is whether we can be persuaded that what we should do is meaningful and that it does not conflict with the basic patterns of our values. Quite often the latter is not conscious but we have the feeling that some action is not compatible without being able to name those values explicitly. In this context the evaluation both of measures as well as the results of the application of measures and rules is of central importance because the compatibility with so-called *"every day values"* concerns our natural ethical moral sensibility. Naturally this can change on the concrete level if, for example, we think of *"impaling our foes on stakes in case of war."*

So the point of all this is to create a moral sensibility. Human beings need to be alerted that what they see or observe is not compatible with their inherent values, which determine their everyday life. Today, of course, something has been reversed and one has the impression that a fulfilled life is not something that can be, but rather something that one can own or that one can consume. One wants to reach fulfillment with the help of luck and other satisfaction and economic goods. Some of us believe that if we do not have any sensibility for our human fellows we are better off. Sensibility and sensitivity are considered as a weakness! One is not strong enough!

We eliminate a very important element of positive empirical values of ethics and indeed its possibility to create areas of reflection and correction. We act as though our economic theories were complete. We think that knowledge can be isolated in a calculator and used without having to know or understand the significance of the rules of the calculator.

Moravcsik Extended - Constraints and the Development of Sensitivity

Taking up the graph in figure 2 above should explain the kind of reasoning essential for our topic in introducing four kinds of background-knowledge areas, i.e. a 4-dimensional semantics as an extension of developments started by Gentzen, Beth and others leading to situation semantics, etc.

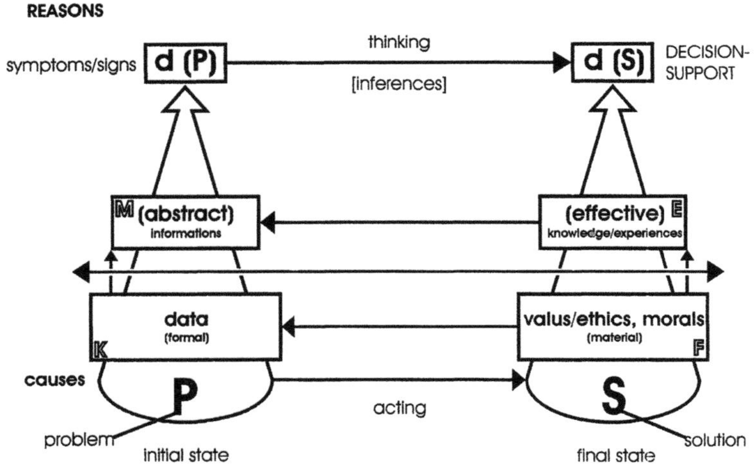

Fig. 4: Means of Analysis (F: = folk knowledge, communal knowledge // E: = experts knowledge, experience // K: = calculi, rules routines checklists // M: = meta-knowledge, knowledge by models, explanatory knowledge)

Let us immediately start with an analysis and synopsis of three examples:

Consider a patient approaching a certain physician, a renowned specialist, to examine his lungs. Looking at the graph [Fig 5] you might consider the situation in the following way: the patient is in state P/medical problem situation and gets his lungs examined, i.e. have them x-rayed. The specialist physician looks at the x-ray picture in d (P). So there will be started a medical initiative, including some means (therapy) to help the patient and to reach a stage or a state S (of the lungs) to be considered as improvement (of the original state P) and showing up in a new x-ray picture d(S) (output) in the future, a picture which would be understood as (possibly positive) outcome S. What is it that links up the connection between d (P) & d (S)?

What links it up is definitely the experience and sensitive interpretation of the patient's symptoms by the physician. But in order to create a good pre-selection [d(S)] of all those x-ray pictures available to the doctor, he may develop a *"set of rules"* or techniques and checklists to be able to reproduce approved of standard results.

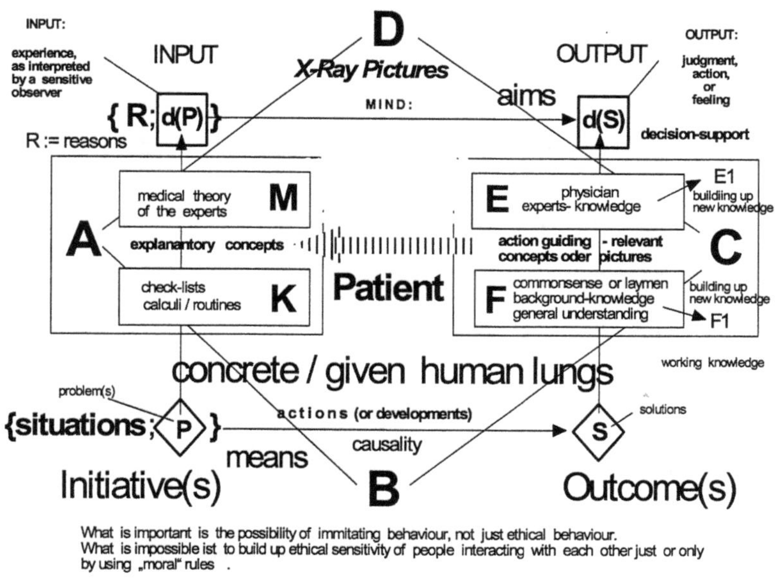

Fig. 5: X-rays and Ethical Argumentation (Moravcsik's idea of the influence of ethical considerations – cf. top level)

These rules (in the graph represented by the field K) may then be used by loosely trained (motivated) assistants using a weak i.e. (with respect to an understanding of x-ray pictures) more or less laymen's background knowledge F, i.e. knowledge of application in daily life.

So the triangle E / K / F analyzes and depicts the situation. The meaning /significance/importance of M [explanatory/meta-knowledge/reflective & corrective knowledge] will be discussed in Fig. 5.

The analysis of the x-ray situation would also hold true of a situation where an experienced officer in E (i.e. with effective, both explanatory and operative knowledge, (knowledge that is both action guiding and descriptive) would instruct his people to do things, obey orders (background knowledge K) and realize actions on the basis of their own background knowledge F. The problem is how much background knowledge, experience, sensibility, responsibility do soldiers need in order to be effective in reaching the situation S aimed at in the prescribed and analyzed description d (S).

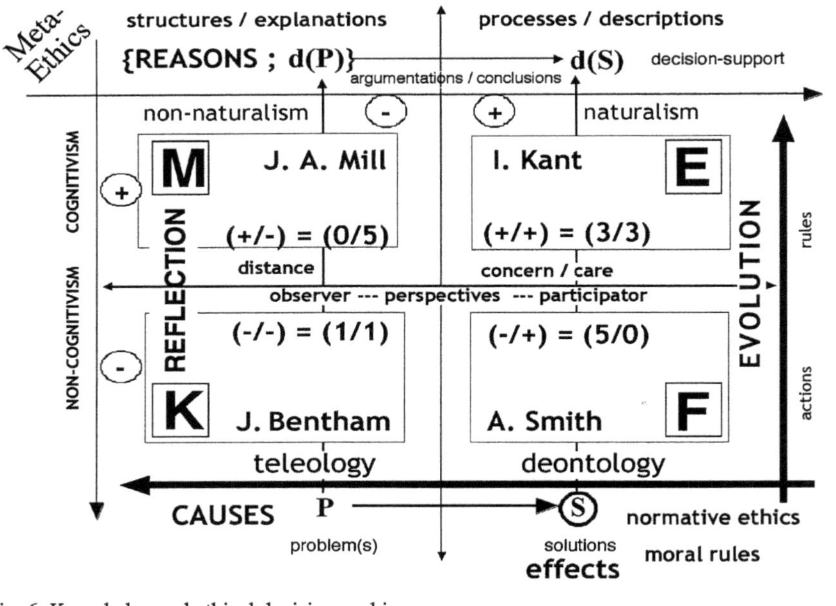

Fig. 6: Knowledge and ethical decision-making

The basic model theory or conception is explained now: Language, Information, and Reality - possibilities of communicating facts, knowledge, feelings, and *"experience."*

> *"Communication between you and me relies on assumptions, associations, commonalities and the kind of agreed shorthand, which no-one could precisely define but which everyone would admit exists. That is one reason why it is an effort to have a proper conversation in a foreign language. Even if I am quite fluent, even if I understand the dictionary definitions of words and phrases, I cannot rely on shorthand with the other party, whose habit of mind is subtly different from my own. Nevertheless, all of us know of times when we have not been able to communicate in words a deep emotion and yet we know we have been understood."* (Winterson, 1995: 79-80)

The following original scheme is a simplified meta-representation of communication unifying linguistic and non-linguistic elements, which above all take into account the coming about of an understanding through the interpretation of signs via different components of background knowledge and which considers the dynamics of the conveyance of knowledge and changes of meaning. 'Knowledge' (e.g. implicit knowledge) results from the mutual relationships of the different components of background knowledge. 'Knowledge' reveals itself in the handling of knowledge. 'Knowledge' emerges through the relations of things to each other. 'Knowledge' mediates between language and reality, defines the handling of linguistically encoded information and determines the relations of language onto reality.

If we communicate knowledge (i.e. belief systems which determine the acceptance or persuasive power of ethical arguments), we have to consider the background knowledge of the recipient in its multiplicity (cf. the components E, F, K, M in the model theoretic scheme). If we want to communicate the transition of a state P into a new state S = Q (in the world, in an attitude, in understanding, in knowledge) or want to make it explicit or even create it (in the recipient), we have to be clear about the means of representation R (e.g. some language) to be used and we also have to clarify which components of background knowledge relate the signs in R to sections of the world W. The transition from P to S = Q is reflected linguistically (in an argument) and therefore also in communicating the acceptance of the transition of □ = d(P) to □ = d(S), i.e. it is reflected in the admission of the relationship of those signs which are assigned to the (more or less real) state-transitions P and Q = S in the realm of representation D. This acceptance in the realm of representation can be strengthened by a deliberate change of relevant components of the background knowledge responsible, in the last resort, for approval and the endowment with meaning. Whether we actually accept and therefore successfully communicate knowledge (especially when dealing with creating and conveying new views, frames of reference, etc.) depends on the interplay of the respective components of our background knowledge.

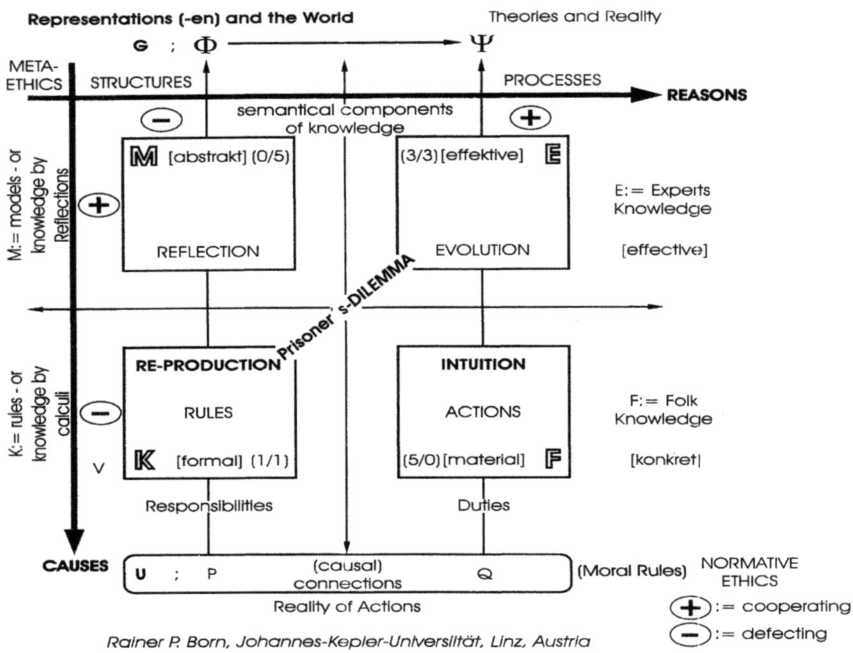

Fig. 7: Communication and ethical decision-making

Here, the relationship between theoretical knowledge T (selected general knowledge A, cf. the left side of x-axis) and vernacular knowledge V (communal or common sense knowledge C, cf. the right side of x-axis) is decisive, since it determines the fine-tuning of new and old knowledge in the concretely chosen area B [= Bereich, cf. R. Carnap] (as section of world/reality, lower part of y-axis) and the representation D (as specially chosen representation, upper part of y-axis). Value-judgments or general ethical considerations, human values and aims in handling new *"knowledge"* are accepted and influence the handling of knowledge and information via the background knowledge.

Computer-Poetry & the Drummer's Example versus Deep Blue - or: Causes and Symptoms Mixed Up

In the following, figure 8 will be illustrated with the help of some ironic comments: When I first introduced the basics of LIR as a sort of multidimensional semantics (there are at least four components of meaning to be considered, which in their complexity need not be taken into account all the time) I invented a surrealistic example concerning the producing of poetry with the help of computer programs and what would be needed to fake an understanding of the "meaning" of poems. At the time the example was technically well ahead of actual practice but Max Black thought that it well described the "reality" of the refereeing system in poetry journals. The essential point with respect to the scheme LIR above was an evaluation function from K to M such that structure models in M (in the classical way of Tarskian Semantics – cf. Jane Bridge's Beginning Model Theory) could be used to evaluate the "meaningfulness" of poems in a strictly technical way. In essence, the meaning would not literally be expressed but only somehow coded. This would be in opposition to the referees in a journal who have to evaluate poems by understanding their content, so to speak, and express the meaning. The point was: what would happen if some evaluation program replaced the referees in M? In which way could the explanatory encoding of meaning be used to freeze a certain literary taste of the time (and when would it turn up a revolution by the next generation of poets). If one looks very carefully he will see that the same strategy in the end was successfully used in programming the chess program *"Deep Blue."* But the point is that *"meaning"* (or the evaluation of situations in a game of chess) is frozen which further links up with Gödels point against Turing, i.e. that meanings can evolve (in Gödels case the intimateness with abstract mathematical terms and their application as leading to a new understanding essential for proper application). In the case of *"Deep Blue"* we do not even know what it really is that has been grasped by the evaluation functions. What if we would use Deep Blue as an experts system to train students of chess? In a medical experts system it is quite clear: If students are trained just by the system and have no chance to learn corrective experience from outside, in the end they will be only as good as the program. And if the medical experts system has a success rate of 80% the rest of 20% will somehow get lost. But there is also one more *"ironic"* (and quite realistic) illustration of the scheme, which somehow may be considered as a reductio ad absurdum picture of simply transferring natural science to studying the mind, in this case meaning.

Possible Relations between Language / Representations, Knowledge and the World

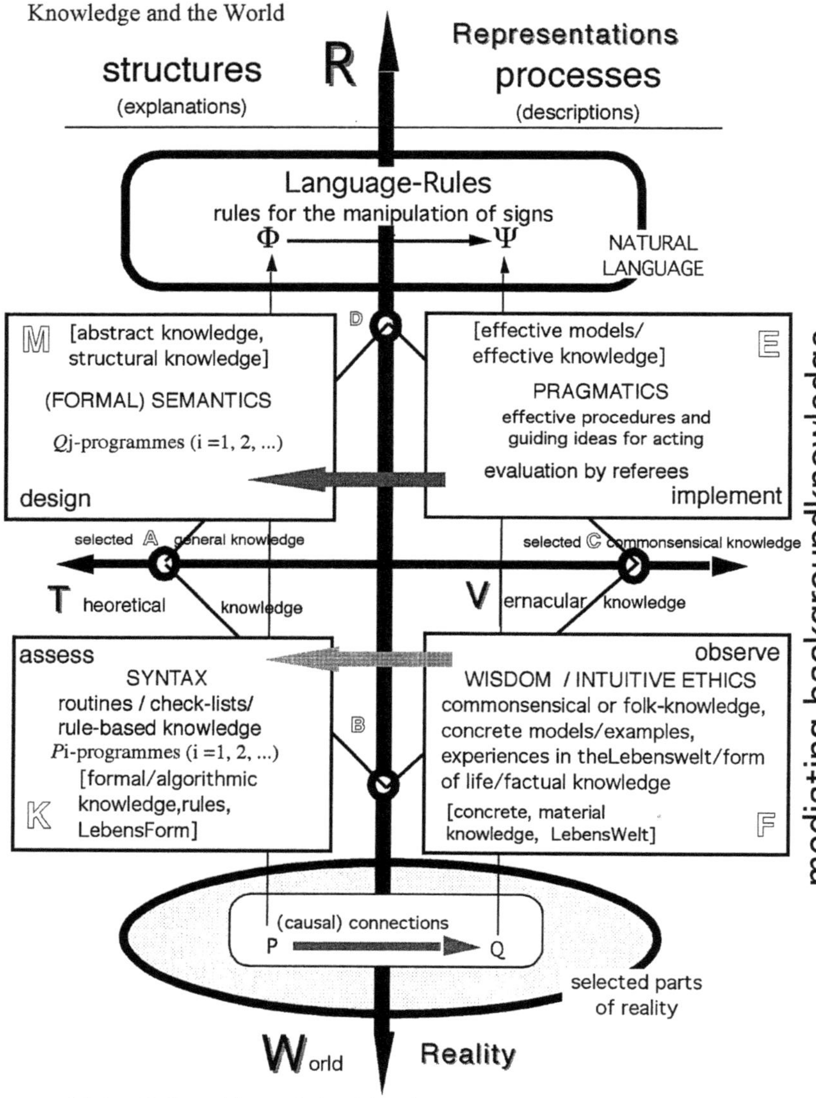

© Rainer P. Born, Johannes-Kepler-Universität, Linz, Austria -- legend in the text

Fig. 8: Possible Relations

The idea is: could we find out the meaning of messages sent by drums without having contact with the people or being able to communicate with them in another way but by drums? The only technical means allowed would be video-cameras and special microphones as well as loudspeakers to fake the noise of drums and watch the reaction behavior. Essential would be the difference between meaning as explicated with the help of M and the meaning (of the drummed noises) as given to the sound by the experts E who use the drums. When do we think we would be justified to say that some model in M has *"grasped"* the meaning of those noises? *"Grasping"* can be used theoretico-explanatory and literally descriptive or operative. I think it is essential to take into account the interplay of the two, i.e. the interplay between theoretical and vernacular (background) knowledge.

The Core: Consequences and Reality

Let us now try to turn the scheme into reality making use of the x-ray example:

Suppose now that in some sort of military context you have to solve a problem where basic ethical attitudes or intuition play a certain role, e.g. the behavior of some soldiers at a *"check-point."* You observe that the soldiers behave in a humanly discriminative way and throw the passport to the feet of some civilians, i.e. they are *"over-doing"* their actual orders (winning the war but losing the peace!).

The aim in such a situation would be to find a solution/instruction (concerning the behavior of the soldiers) whereby they (in some sense even intuitively) understand what they are doing and what the aim of their behavior is. Rules alone will not be enough! Sensitivity for a situation and the application of rules or orders has to be trained, i.e. inherent or implicit ethics concerning what *"should"* be done. Of course creating sensitivity alone is not enough. One has to start with rules in accord with some kind of ethical intuition as a starting point.

Considering what has been done so far we could say/analyze that the codified knowledge of experts ***R*** (E) together with the description of the problem-situation d (P) as it is available to both experts and laymen. Persons with F-background knowledge yield (on the basis of some logic concerning actions) the description d (S) of a solution or state of goal. One hypothesis could be that if we enrich, extend, or combine the background knowledge R (F) with the experts system K(E) - because the rules in K have more or less been explicated by the experts - we could reach more or less the same results d(S) as the experts (think of the x-ray or other specialists). In other words, given d (P) both K & F as well as E will yield d (S) as decision support for some action producing solution/state S if applied to problem-situation/initial-state P.

But looking carefully at K (E) one can easily see that a lot of implicit experience of experts / specialists is contained in K (E) and used to correct misapplications of K. The special experience of the experts which determines the sensitivity of using/applying/dealing also determines the limits of application of K (E) and is missing in F.

When in the beginning we drew the attention to the fact that ethics, in the end, is concerned with judging the consequences of our actions both on the basis of rules as well as duties it was essential to stress that we would have to develop our sensitivity as far as we can go in applying our *"rules"* without overdoing it or overstepping the mark or exhausting our means.

Thus a moral sensitivity has been built up in accord with the *"nature of human beings"* (in education, in a community, in society). In some cases (ethos of the world) this is connected with the hope for a natural ethics of the world. Moral rules could thus be understood that they express our sensibility, responsibility and basic ethical intuition. But what needs to be understood is the fact (an insight going back to Plato and Aristotle - well argued by Moravcsik) that rules by themselves, i.e. without some sensitivity for application and possibility of reflected correction, in the long run are unable to guarantee *"morally acceptable"* action.

Once more it has to be stressed that morally acceptable action is by no means luxury but guarantees the survival of mankind, especially if the cogwheels of rationality (cf. *"rational fools"* by Amartya Sen) run idle and philosophers and scientists stand fast and well secured in the air of their abstractions. In this case, using the scheme again, it is only the triangle F, K, and E which is used to select and determine suitable actions.

But if this insight leads us to conclude that only a dialogue between E and F is essential, a dialogue which establishes the necessary *"sense making"* for the use of routines and schematic knowledge as codified in K, then it might well be that again we are stranded just with a *"receipt"* and we have not understood the *"generalizability"* of the solution (or the possibility of transfer of a solution to other situations in identifying/constructing the essential common structure).

In terms of practice this means that three essential points are missing:

i) The use or applicability of routines K (E) (of expert systems/codified systems of moral rules etc.) needs to be tested in accordance with those people who play the *"role"* of F (i.e. according to their background knowledge or experience).

ii) Laymen and experts have to talk to each other and learn to *"deal"* with K in conjunction, which does not mean that laymen have to build up/learn completely the same knowledge as experts.

iii) Experts have to learn commonsense knowledge/intuitions (e.g. social competence) from the laymen, i.e. they have to enrich their epistemic resolution level as well. (The relation between F and E is not just the one between master and apprentice! Think of *"The Golem ..."* Collins/Pinch - *"Aids"* and laymen versus experts)

The idea that our F-knowledge (even our ethical intuition) needs to be elaborated by schooling and training leads to reconsidering what is really responsible for the success of experts' solutions when the sensitive dealing with routines (to know about the implicitly communicated limits of application) needs to be explained. In this case we have to look from an outside perspective at the matter because otherwise in many cases we will only find out about what experts believe is essential for a solution (or join medical paternalistic terms), what they think that a laymen needs to know, just as doctors sometimes tell us what they believe to know about us, telling us where we should have pain. This fact eventually leads to the construction of structural knowledge (structural models / think of Tarskian semantics and its modern extensions) to codify or explain the knowledge prevailing/residing in E. It means that in M (meta-knowledge, explanatory knowledge which does not need to be literally descriptive or action guiding!) we have to construct a model that in a technical but not descriptive way can be said to *"grasp"* the knowledge of E.

This kind of M-knowledge can be called theoretico-explanatory in contradistinction to the descriptive and operational knowledge in E/F and also K. Thus we have followed up some (single) helix (or just spiral) concerning the development and evolution of knowledge because the knowledge constructed in M [used in arguments as R (M)] can be used to select initiatives to identify knowledge, experience, etc. to enrich F (using an appropriate language and epistemic resolution level) in such a way that in many normal cases F & K will produce solutions close to K & E which means that experts can be liberated from certain standard work and should be able to concentrate on important non-standard matters.

If again we combine this procedure with the ethics matrices (Fig. 5) to look at the influence of evaluation components it means that what matters in F are actions and duties, in K schemata for acting and consequences of actions and in E principles of middle reach [cf. Beauchamp & Childress (2001) Biomedical Ethics], e.g. the categorical imperative of Kant, i.e. regulative ideas, rules and duties (deontology). In the last field M we have to fix the so-called supervising ethical principles that concern both rules and consequences, which are sometimes theoretically-explanatory and will reach their concrete meaning in an effective realm of action.

If one furthermore combines these investigations with considerations concerning the prisoner's dilemma (see appendix) one can correlate the moral action and reflection areas with well known norms from the two-person-game with no repetition (the simplest kind). Certain numbers from the pay-off matrix give an idea of the interplay between cooperation and defection in combination with a general ethical analysis that is essential for an understanding of mankind.

K with its syntactical area of action and with minimal background knowledge corresponds to (1/1). Both agents defect and do not cooperate due to some misunderstanding of the theory behind the prisoner's dilemma. The field F corresponds to (5/0) since a person acting in accordance with duties and therefore cooperates will easily lose - the defector wins. Thinking just in M (knowing explanatory rules) and thinking in consequences will lose against someone who solely thinks in terms of rules and defects if it seems appropriate - therefore (0/5). In E a maximum in cooperation is possible - therefore (3/3).

On the other hand, if one thinks of the *"nature of men,"* one may find that some kind of altruism may have been responsible both for the development and for the survival of mankind. This would mean that our natural or intuitive feelings, our sensitivity and cooperation concerning our fellow humans, constantly needs to be developed and adapted in the area E as an evolutionary dialogue with F and as a reflective analysis via M. The point is that our effective ethical principles formulated in E should only have a middle reach to enable the system to remain flexible and still adaptable.

Thinking back upon the first heart transplantations we remember that we had to change a principle in M and to allow for a different conception of death. In a similar way one can think through the example of enemies and think about which common sense conceptions in F had to be changed and what remained constant in M and in which way E would have to argue differently to install new rules such that enemies should not be. That is, we had to anchor new rules in a changed sensibility in our common sense thinking.

As a conclusion it should be emphasized that we can analyze our procedures and arguments but will find out that our reflections need not be literally action guiding.

The point is rather to choose and explicate our logistics, routines, and instructions in K in such a way that one is able to act *"fast and safe"* in normal situations but is alerted to know how far to go in borderline cases. In justifying as well as in leading (in military contexts) we need to develop a feeling for which principles of selection (thin or rather abstract ethical concepts) need to be identified and explicated in M such that operationalizations or instructions in F and E can both be secured as well as used in a flexible manner. This sounds like squaring a circle. You may remember that the four fields are to be considered as semantics for problem solutions and argumentations of justification in decision support situations. They have different logical status and are not just simple procedures.

F-knowledge in its semantic role is concrete and material. K-knowledge is rather formal. E-knowledge is effective, both descriptive and explanatory, just as the categorical imperative wants to be both.

The same holds true of the understanding of algorithms in a computer program: they want to express both logical and procedural aspects but are also considered (cf. Aristotelian logic) to have causal significance (actually one should distinguish between a conditional and a causal meaning of the linguistic expression *"if, then ..."* and to be aware of those areas where it only seems to be the case that they coincide.)

The last field M concerns the abstract and therefore most general semantic component in the context of logical conclusions. The relation between F/E and K/M is (logically speaking) *"many-one."* That means that *"one"* abstract conception in A = K/M can correspond to or be operationalized by *"many"* possible realizations in C = F/E.

Once more: The proper connection between moral rules and ethical sensibility is inescapable if one wants to develop a flexible system that is able to survive in the long run.

Pragmatic incompleteness

Starting from the fact that our ethical theories are not literally descriptive or action guiding and therefore in a broad sense of meaning are not "complete" we can see that the application of rules is in need of some additional kind of sensibility to be able to correct misapplications.

This kind of incompleteness shows up in some insufficiency of the rules. In some cases it may happen that explanatory conceptions are turned into rules in the wrong way (cf. prisoners' dilemma). From the point of view of philosophy of science one argues that theoretico-explanatory models must not be turned into operationally conceived systems for guiding actions.

If at all we need some possibility of correction, which again calls for a new or re-established kind of sensibility.

Appendix: Prisoner's-Dilemma

Shortcut of two-person prisoner's dilemma without repetition adapted from Hofstadter

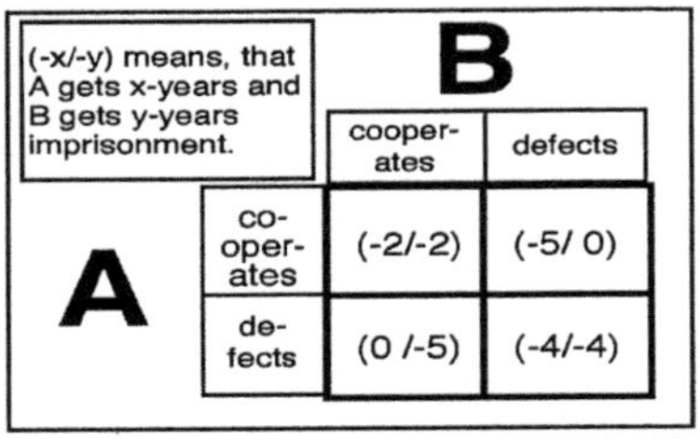

pay-off matrix for real prisoners and their dilemma (adapted from D. Hofstadter)

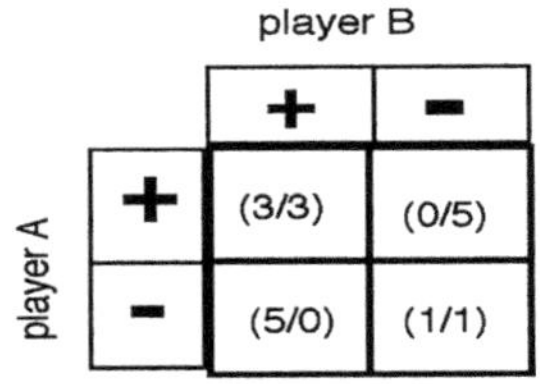

canonical pay-off matrix (adapted from D. Hofstadter)

Short version: A and B have committed a „crime" and were caught. The prosecuting attorney cannot proove much but could send both of them to jail for two years [(-2/-2) in the graphic] (i. e. if none of them confesses). If one of them acts as principal witness he will be free and the other will be imprisoned for 5 yeears [(0/-5) oder (-5/0)]. If both confess both will get 4 years [in the pay-off matrix: (-4/-4)]. -- If one adds 5 everywhere, one gets the general matrix (right hand oicture). "+" means to „cooperate" "-" means to "defect".

The result? An illustration of game theory was turned into an advice for acting!

References:

Beauchamp, Tom L.; Childress, James F. (2001): Principles of Biomedical Ethics. Oxford.

Collins, Harry; Pinch, Trevor (1999): Der Golem der Forschung. Berlin.

Collins, Harry; Pinch, Trevor (2000): Der Golem der Technologie. Berlin.

Davis, Morton (1972): Spieltheorie für Nichtmathematiker. München.

De Geus, Arie (1997): The Living Company. Habits for Survival in a Turbulent Business Environment. Boston.

Fehr, Ernst & Fischbacher, Urs: The nature of human altruism. University of Zurich, Institute for Empirical Research in Economics, Blümlisalpstrasse 10, CH-8006 Zurich, Switzerland.

Gardner, Howard; Csikszentmihalyi; Damon, William (2001): Good Work. When Excellence and Ethics Meet. New York.

Gruen, Arno (2003): Verratene Liebe - Falsche Götter. 'Stuttgart.

Hargreaves, Heap-Shaun (1994): The Theory of Choice. Oxford.

Hofstadter, D. (1998): Tit for Tat. In: Spektrum der Wissenschaft, Digest Kooperation und Konkurrenz 1/1998:60-66.

Michaels, Anne (1997): Fugitive Pieces. London.

Moravcsik, Julius (2003): Was Menschen verbindet. Übersetzt und hrsg. v. Otto Neumaier, St. Augustin (1. Buchveröffentlichung in Deutsch).

Morgenstern, Oskar (1963): Spieltheorie und Wirtschaftswissenchaft, Wien.

Neumann, John; Morgenstern, Oskar (1944): Spieltheorie und wirtschaftliches Verhalten. New York.

Putnam, Hilary (2002): The Collapse of the Fact/Value Dichotomy and other Essays. Harvard.

Shubik, M. (1967): Spieltheorie und Untersuchung des sozialen Verhaltens. New York.

Siegmund, Karl (1996): Spielpläne. München.

Singer, Peter (2002): Wie sollen wir leben? Ethik in einer egoistischen Zeit. München.

Slater, Robert (2003): Im Auge des Sturms. Die Insider-Story von Cisco, John Chambers, dem Börsencrash und dem Wiederaufstieg. Hamburg.

Pierer, Heinrich von; Homann, Karl; Lübbe-Wolff, Gertrude (2003): Zwischen Profit und Moral. Für eine menschliche Wirtschaft. München.

Weick, Karl. E.; Sutcliffe, Kathleen M. (2003): Das Unerwartete managen. Wie Unternehmen aus Extremsituationen lernen. Stuttgart.

Winterson, J. (1995): Art Objects - A Veil of Words - With reference to the Waves. London. 79-80.

Hubert ANNEN, Military Academy at the ETH Zurich, Zurich

ACTION RESEARCH 1:1 EXEMPLIFIED BY THE PROJECT "SUPPORT-ORIENTED APPRAISAL"

In the contribution *"Action Research as a method for scientific thinking and acting in military pedagogy"* it was illustrated which characteristics define action research and how action research differs from other methods. Against this background it can be concluded that this procedure is especially well suited to research and action fields of military pedagogy and, therefore, should be carried out more often (Annen, 2002). This contribution actually refers to the project *"Support-oriented appraisal"* in which a new appraisal system for military Non-commissioned officers (NCO) and officers based on strict observance of the principles of action research in cooperation with experienced people was developed (Annen, 2000). The present paper explains this project with the aim to illustrate with a concrete example the typical sequence *"Look - Think – Act"* (Stringer, 1996), the specific quality factors, as well as chances and dangers of action research.

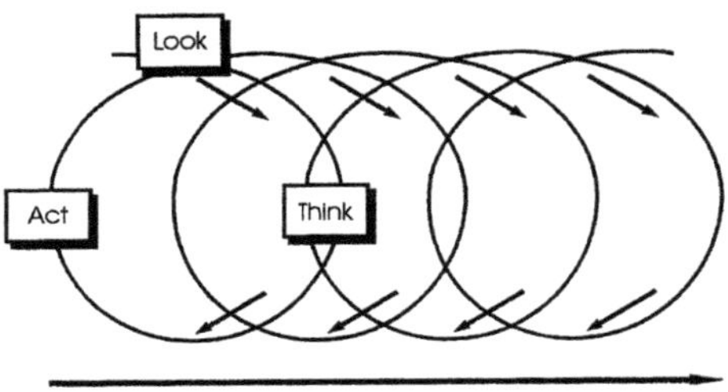

Starting Position and Issues

At first, it might be difficult to see a direct link between personnel appraisal and military pedagogy. But when looking at certain aspects of the issue this link becomes quite obvious. The appraisal includes an educational aspect in so far as the superior appraises behavior and achievements and gives concrete feedback and thus communicates what he or she would like to have improved, kept, or even eliminated. This is to a large extent in accordance with Brezinka's (1990) definition of education. It is at the appraisal interview that superior and subordinate must come into personal contact. At this occasion, the relational aspect and with it the educational aspect (Nohl, 1933) becomes explicitly manifest. Furthermore, the interface with pedagogy becomes evident because, as stated in the *"classical"* definition of the role of a teacher by the Deutsche Bildungsrat, appraisal is one of the core functions of a teacher. (Gudjons, 1999) The same idea is expressed in the job regulations by the Swiss Teachers Association, which highlights the importance of appraisal as part of their classroom work. (LCH, 1999) Therefore it is obvious that appraising or qualifying subordinates in a training army such as the Swiss Armed Forces is an important means of leadership and education and, thus, a major focus of military pedagogy.

The fact that the chances of this instrument are widely underestimated and that appraisal is often considered an administrative chore was one of the starting points of this project. It should not be forgotten that via this instrument our young militia NCOs and officers receive targeted feedback on their leadership activities and that by carrying out appraisals themselves they can gain valuable experience with an important leadership tool, which is not without profit for their civilian career. Shaping a leadership culture is a second important aspect. According to various sources it has been demonstrated that the tool of appraisal is one of the few tangible means to influence leadership culture. (Schein, 1995; Hoenle, 1996) Since 1991 the textbook *"Human-oriented Leadership (Menschenorientierte Führung)"* (Steiger, 1990) has been the official leadership handbook in the Swiss Armed Forces. Discussing these principles in courses is essential - but not enough. As soon as the appraisal criteria for military commanders are more clearly focused on human-oriented leadership, the commitment will increase. As a third point it must be mentioned that the appraisal system for military cadres has undergone only slight changes since the beginning of the seventies. And although it had been the object of various scientific research projects, the well founded suggestions to change the system have shown little or no effect.

Hence three questions can be asked:

1. What should a modern qualification system for militia cadres look like in order to contribute to a leadership culture and to the development of leadership personality?
2. How can the quality of an appraisal system be assessed and how good is the new support-oriented appraisal instrument?
3. Which factors play a role during the process of change?

Action Research

The questions asked above include the ambition to produce longer lasting results than the former studies which dealt with the same research topic. The method that imposes itself is action research, which explicitly aims at the process of implementing planned modifications. The researcher leaves his ivory tower and becomes an acting person. But during certain phases he again moves away from that function and concentrates on scientific reflection. The persons in the field are not only confronted with the product when it is finished - shelving would then be inevitable - but they are involved in the process right from the start. In order to meet scientific criteria, the procedure and the (intermediate) results have to be continuously documented under various perspectives. Thus, the basis for the ultimate requirement of the action research oriented procedure is created: It must result in a contribution to the scientific discussion. The action researcher is entrusted with the task to preserve the link between the scientific and the practical system and therefore to bridge the almost unbridgeable gap between the two worlds. The following chapter will illustrate the implications for the project concerned.

Procedure for the Project "Support-Oriented Appraisal"

The procedure follows the sequence *"Look - Think – Act."* I will not give a detailed account of the (intermediate) results. The main goal is to show what consequences had been drawn regarding the further procedure.

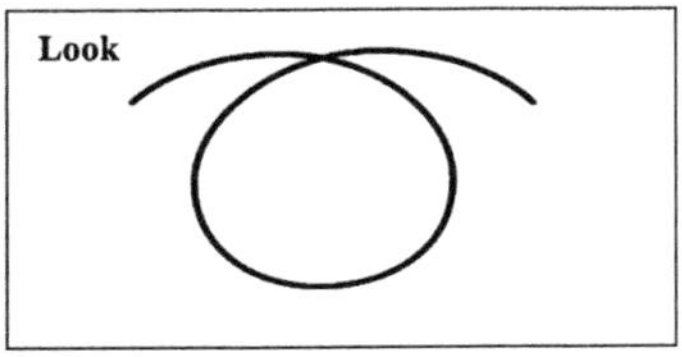

Establishing the facts

- analysis of documents
- written survey
- semi-structured interview

The current appraisal system is well accepted by those involved. People welcome an enhancement, but without expressing an urgent need for something new. Some work will therefore have to be done to convince people to introduce and apply this new procedure. Leadership is often differently perceived than it was meant. Consequently, people who carry out the appraisal assess the procedure in a significantly more positive way than those who are appraised. Because of this perception a better transparency and a higher participation in the appraisal process is needed. When it comes to the 360°-feedback and the importance of the appraisal meeting, the fear of a time-consuming procedure and of a partial limitation of the power for the so far single *"appraiser"* becomes evident. Therefore a practical tool has to be developed, which clearly shows the advantages of additional appraisal perspectives.

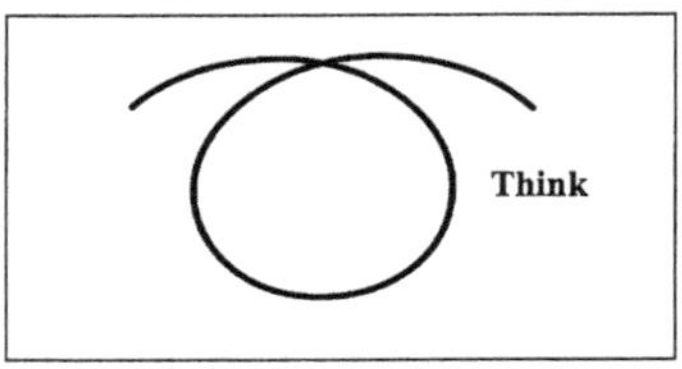

Planning and development

- group discussions
- analysis of documents
- member check

A new qualification system was created in cooperation with professional cadres of five military bases. The result was an open appraisal form which clearly distinguishes between behavioral dimensions and measurable achievements. The person who carries out the appraisal has to set and assess concrete examples of behavior. Additionally there is enough space left to list hints that would support the desired behavior. The subordinate fills in the same form as self-appraisal. At the appraisal meeting, the two sides compare their notes and the hints for support will be transformed into concrete measures. Such forms and aids were developed by the researcher and submitted to the practitioners *"for approval."*

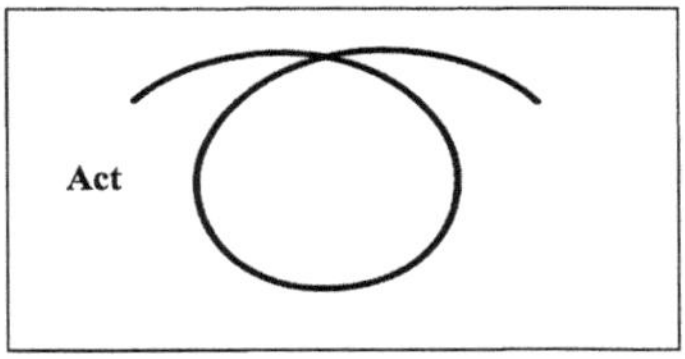

First round

- instruction / evaluation meeting
- statistical data
- written survey

This procedure was used and evaluated in the officers' and recruits' schools which took part in the test. It was confirmed that self-appraisal is not of lesser quality than appraisal carried out by the superior. The data showed a certain self-criticism and an astonishingly stable self-image. Talks with the users and questionnaire furnished information concerning social validity. (Schuler, 1998) The feedback was overwhelmingly positive, but not better than former procedures. The form and the instruction were well accepted. The appraisal meeting went well according to the persons involved. However, the chances of this enlarged instrument were only partially recognized and the procedure was considered to be time-consuming.

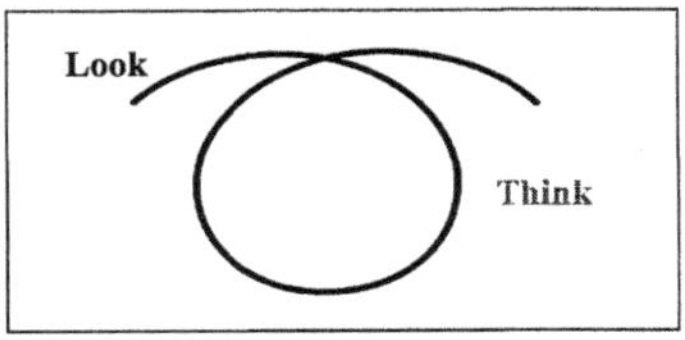

Reflection

- analysis of documents
- member check

At the ensuing discussion with the professional NCOs and officers, the reduction of the time needed for this procedure was a central issue. As a result, documentation with suggestions of formulations was added, and the part with the hints for support was simplified.

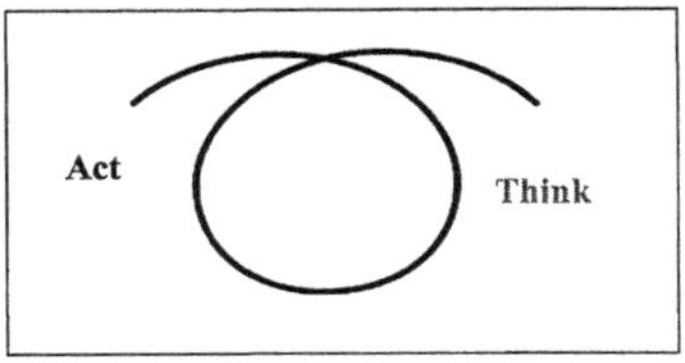

Second round

- instruction / evaluation meetings
- statistical data
- written survey

The feedbacks by those involved were similar to the first round: They were mostly positive, appreciated the chances of self-appraisal and mentioned favorably the frankness and honesty at the appraisal interviews. The efforts made in order to reduce time did not prove effective, as it was still thought that the new procedure would take up too much time.

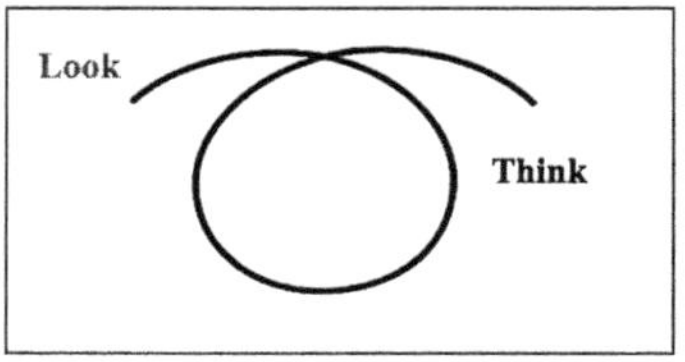

Reflection and outcome

- analysis of documents
- member check
- research report

In order to reduce the time load substantially, each behavioral dimension in the form was illustrated with concrete examples, which were generated from the qualitative assessment of the forms filled in so far (more than 1000). Thus, a heavy load of writing was taken off the appraisers which were welcomed by the persons involved. Time load was not the big problem any more, and the positive elements of the new system were better noticed. By the same token, the content of the form can be considered as a job profile for the Swiss officers.

The *"support-oriented appraisal"* has not only become an integral part of the schools who took part in the test phase, at the initiative of the school commanders it was also introduced at three more military bases. At the moment, detailed instructional material is being produced and it seems as if the new manual for qualification in the Swiss Armed Forces, which came in force in 2004, will integrate the core elements of *"support-oriented appraisal."* Detailed evaluation results have already been published and are thus open for scientific discussion. As the aim of this paper is to illustrate the effectiveness of action research oriented procedure, only findings pertaining to that field will be discussed in the following.

Factors of the Process of Change

Processes of change contain some fundamental characteristics whose consideration has proved useful for similar projects:

- The highest level of command has to be favorable to the project, so that the researcher can get the necessary leeway.
- The practitioners at the front must be willing to actively cooperate and put in their thought (commitment). They must be willing to spend extra time and to critically represent the view of the user.
- Involving the users must not be a pro forma thing. Only those who are taken seriously and whose contributions are taken into consideration will identify themselves fully with the project. Thus conditions are created, which will make the project self-sufficient.
- In order to bridge the gap between scientific and practical system, the researcher needs good communication and negotiation skills, he has to be

familiar with the environment and persevering enough to keep the project running.

- The principles and guidelines of action research are a valuable aid when obstacles on the side of the users tend to distract the view from the full picture.

Conclusion

The requirements for the action researcher can be applied to the military pedagogue. The military pedagogue constantly has to test his findings or theses against practice, at the same time he has to convince the practitioner to reflect his daily thinking and acting, his thought patterns, and routines. Experience shows that one tends to be absorbed too fast by practical problems and questions. Scientific distance and critical reflection have to be sought consciously, even if more apparent tasks and needs seem to be more important. The pedagogue can disturb, question familiar things, and bring repressed thoughts to the surface. Once he succeeds in showing the practitioners the importance of such mechanisms, he has achieved the conditions necessary for them to meet the complex requirements of a constantly changing world.

References

Annen, H. (2000): Förderwirksame Beurteilung. Frauenfeld, Huber.

Annen, H. (2002): Action research as a method for scientific thinking and acting in military pedagogy. In: H. Florian (eds.): Military pedagogy - an international survey. Frankfurt a.M., Peter Lang.

Brezinka, W. (1990): Grundbegriffe der Erziehungswissenschaft. 5. neubearb. Aufl., München, Reinhardt.

Gudjons, H. (1999): Pädagogisches Grundwissen. 6. erg. Aufl., Bad Heilbrunn, Klinkhardt.

Hoenle, S. (1995): Führungskultur in der Schweizer Armee. Frauenfeld, Huber.

LCH - Dachverband Schweizer Lehrerinnen und Lehrer (1999): LCH Standesregeln. Verfügbar unter: www.lch.ch/standesregeln.htm, Juli 2002.

Nohl, H. (1933): Der pädagogische Bezug und die Bildungsgemeinschaft. In: Kluge, N. (eds.) (1973): Das pädagogische Verhältnis. Darmstadt, Wissenschaftliche Buchgesellschaft.

Schein, E.H. (1995): Unternehmenskultur. Frankfurt a.M., Campus.

Schuler, H. (1998): Psychologische Personalauswahl. Göttingen; Bern; Toronto; Seattle, Hogrefe.

Steiger, R. (1991): Menschenorientierte Führung. Lehrschrift 51.307, Frauenfeld, Huber.

Stringer, E.T. (1996): Action Research - A Handbook for Practitioners. Thousand Oaks; London; New Delhi, Sage Publications.

Jürgen OELKERS, University of Zurich, Zurich

GERMAN PEDAGOGY AND THE QUESTION OF MILITARY EDUCATION IN THE TWENTIETH CENTURY *)

Determination of the Problem

In 1897 German professor of pedagogy Wilhelm Rein[29] put forward - in the fourth volume of the then authoritative *Encyklopädisches Handbuch der Pädagogik*[30] [Encyclopedic Manual of Pedagogy] under his editorship - that militarism and school education were incompatible and stood in contradiction to one another. By *"militarism,"* Rein did not mean the expression of military life in the German Empire, but rather the entire area of the military, including *"military education."* Military education and school education, wrote Rein, must be strictly differentiated, just as the school could not be identical to the army and its organization. Pedagogy applied to a different world than the military world. The main reason for this dichotomy arose from their differing goals: military education was to be understood as the training of the soldier for the army, while school education, in contrast, was the fostering of the individual (Rein, 1897: 771).

Rein wrote specifically on military education as follows:

> *"Without doubt, the ultimate goal of army education when working on the body and mind of the soldier, is its mass effect. It aims to make the individual capable of taking on and fulfilling his role as a member serving the whole. In the military organization, the meaning of the individual consists in foregoing all expression of one's own will and fitting oneself as a willing part into the whole, so that the whole may work like an artificially constructed machine, whose parts - meshing as do gears - work together to achieve the goal. Militarism views human beings not as individual beings, who are meant to apply to particular spheres of activities their corresponding abilities and*

---

*) All quotations from German sources are translated by the author.

29 Wilhelm Rein (1847-1929) first studied theology in Jena and received a doctorate in Leipzig with a dissertation on the pedagogy of J.F. Herbart. After teaching primary school, Rein entered the field of teacher education. He taught at the Teachers' Seminary in Weimer and then served as director of the Teachers' Seminary in Eisenach from 1876. In 1886 he was appointed Honorary Professor of Pedagogy at the University of Jena and represented pedagogy there until 1923. He became an internationally influential educational theorist with his "scientific pedagogy" based on Herbart's philosophy and pedagogy. By the end of the nineteenth century, Rein was one of the leading, if not *the* leading, professors of pedagogy in the German-speaking world.

30 Under the headword "*Militarismus und Schulerziehung.*" (Rein, 1897: 771-774)

propensities, but rather as material, like that ordered to be worked upon by the master builder, until it appears to be properly prepared to be inserted in a particular position in the building structure as a whole." (ibid.:772/773)

Rein's comments were intended as criticism of two other articles that appeared in the same encyclopedia; one article dealt generally with the military educational system, and the other discussed the advantages of military youth games.[31] For Rein, application of these areas to school and general education was a transgression, an unacceptable conceptual crossing of the lines that must be rejected. The one, education, should have nothing to do with the other, the military.

As it followed, however, German pedagogy more and more moved away from this differentiation. The connection between the military and education became ever closer, both in theory and practice, from the time during the early German Empire when *Erziehung zur Wehrhaftigkeit*[32] (Storenburg, 1878) was a focus of interest. At the turn of the twentieth century *Wehrkraft durch Erziehung*[33] (von Schenckendorff & Lorenz, 1904) also became the topos for the development of the state schools, which culminated in considering *"teachers [Lehrstand]"* to be *"warriors [Wehrstand]."* (Willmann, 1914) Military education was not infrequently equated with education altogether, and the militarization of society was seen as educational progress.

If we look at the uninhibited and self-understood way in which the *"arming"* of education was conducted, its widespread acceptance, and the illustrious authors who participated in the discussion, then Rein's warning about transgressing boundaries appears positively clairvoyant. After 1900, German pedagogy became massively involved in the issues of not only military education, but also war education. Alternative positions[34] were ignored or excluded, and this with the participation of all the leading pedagogical authors. After 1914 particularly, the result was a phalanx that would persist up to 1945.

The underlying theoretical assumption was that the *Volk*, the people, could be educated as a unit, an integrated whole [*Einheit*]. The target group of military education was not merely future soldiers, but the entirety of the *Volk*. The aim was not merely social integration, but rather conformity in action and thought.

---

31 The headwords of the two articles were "*Militär-Erziehungs- und Bildungswesen*" (Rein 1897:757-767) and "*Militärische Jugendspiele*" (ibid.:767-771). The first article was written by von Scharfenort, who between 1859 and 1892 had published a history of the Prussian Cadet Corps in several volumes.

32 Education towards the ability to defend oneself militarily.

33 Military Strength through Education.

34 Such as Zollinger (1894). The most well-known warning about war and war education is Wilhelm Lamszus' *Das Menschenschlachthaus* of 1913.

The *Volk* itself was to become *"fortified,"* which meant that as early as 1871, army and school were conceived in such a way that they appeared to be analogous principles. (von Rhöden, 1871) From there it was but a short step to conceptualize education of the *Volk* in accordance with the organization of the *Heer*, or army, thereby replacing the pedagogical principle with the military principle.

People, Army, and Education

It posed no problem to the influential German pedagogue and school reformer Berthold Otto[35] to put forward the German army as the *"best organized community"* and to put the case for the individual as *"part/member of the whole."* Incorporation of the individual [*Eingliederung*] along the model of the army was to be the foundation of political organization of the people. In this sense, education was to be understood in principle in a military meaning.

> *"The great task of forming, from hundreds of thousands of people, one organic body, filled with o n e spirit and will, into which all members fuse and mesh as do only the limbs and members of the individual physical body so completely, this great task, since the world began, has never been so completely fulfilled as in the German Army."* (Otto 1914: 164)

This notion treating the masses and the army as equivalent and viewing this, moreover, as an *"expression of the deepest will of the people"* (ibid: 168) was not rescinded in the wake of the casualties of the First World War. On the contrary, the idea became ever stronger, after Otto had already in 1905 encountered no obstacles in using the opportunity presented by an excursion to try out his highly praised method of *"holistic teaching"* in Kiel, the harbor of the German Navy.[36] The army was seen as an educationally exemplary organization, the role model that the school of the future would have to follow. The school's main task was education towards obedience as prerequisite for military service:

---

[35] Berthold Otto (1859-1933) studied a number of fields in Kiel and Berlin, but did not complete a degree. He taught as a private tutor from 1883 to 1887 and then worked as a night editor in Hamburg. From 1890-1902 Otto was an editor at Brockhaus-Lexikon in Leipzig. In 1901 he founded the journal *Der Hauslehrer - Wochenschrift für den geistigen Verkehr mit Kindern* [The Private Tutor - Weekly Journal for Intellectual Exchange with Children]. In 1902 the Prussian Ministry for Culture offered Otto a position in Berlin as freelance writer and school reformer, guaranteeing his previous salary and independence from the authorities. Otto founded the *Hauslehrerschule* [Private Tutor School] that brought him fame in Lichterfelde near Berlin in 1906. He headed the school until 1933. There have been no recent studies of Otto and his work.

[36] A brochure by Otto describing the excursion (Otto, 1906) had been published in 18 editions by 1911.

> *"The school of the future ... supplies soldiers who do not need to learn obedience after entering the army, but who from the very start are not only convinced of the necessity for obedience, but have a complete - I would like to say scientific - understanding of this necessity. We do not educate towards b l i n d, but rather to o p e n - e y e d obedience, and therefore to a b s o l u t e obedience."* (ibid: 174)

In 1918 Otto utilized *"daily reports from the Front"* to point out that Rein's dichotomy was non-existent. Military and education were not contradictions, but rather equivalents, for in both areas, individuality was needed in addition to incorporation. The *"heroes"* of war had to have a degree of *"independence"* at their disposal that would far surpass that of any entrepreneur. (Otto, 1918: 163f) In this sense, Otto is even viewing militarism as a German alternative to the free market, as a negation of the *invisible hand*, which in the theories of English liberalism was seen to regulate the market.[37]

> *"Militarism does not at all rely on spontaneous creation of order in social existence. On the contrary, it takes the uttermost pains to produce the true great order - that undisrupted cooperation in large things and small - through daily, laborious effort as well as on a thorough, working scientific basis. And it is here in successful to a degree that, as this Great War has shown, borders on enormity. It succeeds ... without in any way reducing the independence of character or mind."*[38] (ibid: 167)

It was *"a fundamental military law"* that the *Führer* can demand nothing *"that the soldier is not willing to do. From the very start, the soldier goes to war with the firm resolution to sacrifice his life if necessary."* (ibid: 195) The soldier finds his reward in *"the joy of acting"* or the *"joy of succeeding"* - such motivations should be taken advantage of in order to build a *"militaristic national economy."* (ibid: 198)

[37] This goes back to Werner Sombart (1915), who argued that there were two types of human being, the English *Trader* and the German *Hero*, the trader being dominated by the laws of the market economy and accordingly lacking in moral character, and the hero incorporated in the people and, accordingly, of high morals and spirituality. What emerged from this was *Deutsche Sozialismus*, introduced in Sombart, 1934, which was a clear declaration of the *Führer* state. Polemic against England was widespread in the German literature of the wartime period. (e.g. Wagner, 1914)

[38] *"As the World War has just illustrated, it only takes an opportunity to come along, and all the world will see how much independence of character, how much independent thinking, what high courage in personal responsibility is shown not only by every military leader, but also by every second-lieutenant, and even by every non-commissioned officer and finally, in every man, and how superbly the military institutions have understood how to bring out these natural gifts of man."* (Otto, 1918:167-168)

In spite of these ideas, Berthold Otto became one of the leading *"reform pedagogues"* (Germany's progressive education movement) after 1918.[39] His *"Private Tutor School"* in Berlin-Lichterfelde was held to be the epitome of modern, child-centered pedagogy. Otto's militaristic opinions were ignored. Indeed, they have not been addressed up to the present day. In particular, they have not been viewed in the context of a widespread trend that can be termed *war pedagogy* and that gained massive importance in the First World War. In no way was Otto articulating an outsider view. On the contrary, his voice was decisive in the forming of the views of the majority.

Military Education, or "War Pedagogy"

During the First World War, many philosophers and practically all of Germany's renowned pedagogues and educational theorists not only strongly supported the war, but were also in favor further intensifying war education. One of the main issues was how a basis could be found for a positive philosophy of war that would stipulate partisanship for the German war. In connection with this discussion, it was left to Max Scheler[40] to bring the concept of the *"Genius of War"* [*Genius des Krieges*] into play,[41] whereby war decisively determines the future of the society.

> *"War is not a 'fight' whose outcome is decided by rules. It is a function of a concrete, u n i q u e process of nations' and states' growth and becoming and determines the possibility, extent, and direction of all further growth; it is no mere dependent function of their present and past state of being. For the sake of the future, not insofar as it is regulated by law and "calculable," but rather insofar as it can indeed be shaped by free actions, wars have been conducted*

---

[39] The leading journal of German reform pedagogy, *Der Säemann*, in its edition of August 1914 contained a picture that depicts Berthold Otto clearly as a pedagogical leader (*Der Säemann*, 1914:280). Herman Nohl called Otto the *"most original of our German pedagogues"* as late as 1960. (Nohl, 1960:15)

[40] Max Scheler (1874-1928) studied philosophy and sociology in Berlin, where he received a doctorate in 1897. He completed his habilitation under Rudolf Eucken in Jena in 1899. After teaching in Jena, Scheler became a member of the phenomenological circle around central figures like Theodor Lipps, Moritz Geiger and Alexander Pfänder in Munich in 1907. Scheler lectured to the Göttinger Philosophical Society in 1910 and 1911. He was drafted into the army in 1914, but was exempted from military service due to poor eyesight. In 1919 he became a professor of philosophy and sociology at the newly founded University of Cologne. His war writings from 1914 to 1918 are largely ignored.

[41] The title of Scheler's book is an allusion to the third chapter of the first volume of Clausewitz' 1832 *Vom Kriege*. Clausewitz (1980: 52) did not use the term *Genius des Krieges* but wrote instead of the *kriegerischen Genius* (military genius), by which he meant the common orientation of all the mental and spiritual powers, particularly the power of rational thinking, towards the events of war.

in all eras. It is in war that being is 'created' in the first place, a being that all international agreements must take as a given in order to have meaning. It is the boundary and the source, not some kind of mere 'fight'." (Scheler, 1917: 19)

The roots of war, Scheler goes on, lie *"in the very nature of life itself,"* (ibid: 35) although we cannot account for life on the basis of biology alone. What stands behind war is not a Darwinistic *"struggle for survival,"* (ibid: 38) but rather *"the deeper principle that is closer to the essence of life - the unspoiled rise in power of the nobler and higher-natured human groups as they expand their environments and have an effect on their worlds."* (ibid: 41) Along the lines of this principle, *rising in power* lies *"in the very essence of the state itself,"* and the warring state is the *"state in the highest realization of its existence."* (ibid: 42)

"The state is, in the first place, a personhood of mental will [geistige Willenspersönlichkeit], supported by the vital organism of the people. Neither one is any less real than the individual organisms and persons; however, it is the mental principle that guides and steers the vital substructure. War is, accordingly, at the same time both an expressive behavior - an impulsive, thrust-like discharge of this indivisible and only in abstracto divisible whole that is the 'mental-vital nation' - and the deliberate steering of this discharge, for a particular national purpose, in order to impose the will of the state on another state by force."[42] (ibid: 56)

The terms *"expressive behaviors"* [*Gebärden des Ausdrucks*] and expressive *"discharge"* [*Entladung*] of instincts can be traced back to Wilhelm Worringer's theory of experience [*Erleben*] and seen in connection with numerous attempts to aestheticize war. The best known German example is Ernst Jünger's novel *In Stahlgewittern* of 1920.[43] For art history, Worringer distinguished between *abstraction* and *empathy* as two poles of man's sensitivity, connected by a highly effective psychological assumption that lies between expressionism and Ernst Jünger. Worringer states that it is from man's elementary experience that

[42] Scheler's main thesis goes back to works such as Schmidt-Gibichenfels (1912) essay on the *"cultural factor"* war as the *"creator and upholder of states."* Otto Schmidt-Gibichenfels (born in 1861) was editor of the racist journal *Politisch-Anthropologische Monatsschrift* [Political-Anthropological Monthly Journal], founded in 1902 and published in Berlin-Steglitz. Schmidt-Gibichenfels' 1912 essay was published again in 1913 in a special edition of over 10,000 copies by the *Vaterländischen Gesellschaft zur Verbreitung von Geschichtskenntnissen* [Fatherland Society for the Dissemination of Historical Knowledge] and distributed to political decision-makers.

[43] Narratives by war poets, such as *Kriegsdichter erzählen* (Velmede, 1937), is a genre in its own right. It was developed intensively after 1918 and again after 1933.

artistic abstraction arises, which does not require knowledge, but rather experience alone. Man has a strong impulse towards abstraction, which *"without mediation by the intellect"* creates forms directly. (Worringer, 1959: 52ff) In this way, the *Front experience* (Fronterlebnis) can be seen as the condition that gives rise to authentic literature, without allowing intellectual distance:

> *"Here iron and blood fertilize the spirit* [Geist] *in a dark manner, and the secret of 'rebirth' encompasses not only the state, which in every war arises anew, but also the p o w e r s which swell behind the state and the completed culture in their ever-new production."* (Scheler, 1917: 65)

The pedagogization of war was established at a lower level. Generally, it was common to link the *"Psychology of Great Military Leaders"* (Ziehen, 1916) with an educational role model function, so that already in 1892 Felix Dahn could present *Moltke as Educator* to the public.[44] It was also common for former officers to discuss education towards national pride and honor and to bring *German Race Politics* (Meinhold, 1908) into the mix.[45] Captains in the Medical Corps were highly supported in their praise of the educational value of the German Boy Scouts and the Wehrkraft-movement. (Lion, 1913) Higher ranking officers suggested long before the outbreak of the war the possibility of a *World War* that would transcend local boundaries and for which inner preparation would be necessary (von Bruchhausen, 1906). This dimension not least explains why *"war education"* could become a dominant pedagogical topic.

In the First World War, one of the leading German school reformers, Hugo Gaudig[46] saw the purpose of the *"German school"* in promoting and fortifying the *"national consciousness of unity,"* which the war itself had created, as well

---

[44] Written originally in 1891 as the introduction to Moltke's work on the Prussian-French War of 1870/1871, Dahn's essay appeared later as a separate imprint. Helmuth Graf von Moltke (1800-1891) was Head of General Staff of the Prussian Army from 1858 to 1888, and in this capacity, he was responsible for the three Prussian wars between 1864 and 1871 (Bucholz, 2001). As a hero, he would later be presented as a role model and find pedagogical admiration.

[45] Captain Eberhard Meinhold referred mainly to Otto Ammons work *Die Gesellschaftsordnung und ihre natürlichen Grundlagen* [Social Order and its Natural Foundations] (1895).

[46] Hugo Gaudig (1860-1923) studied theology and philology at the University of Halle and received a doctorate in theology in 1883. In 1884 he completed the probationary year for higher teaching at the *Realgymnasium* in Gera; in 1896 he was appointed as director of the girls' higher school and the Teaching Seminary of the Francke Foundations. In 1900 Gaudig became the director of the girls' higher school and the Teaching Seminary in Leipzig. Here he developed the concept of the *"work school"* [*Arbeitsschule*] in the sense of active learning, which became one of the central concepts guiding German school reform before and after the First World War (compare Oelkers, 2004).

as the risk that after the war, this consciousness could be thoughtlessly squandered. (Gaudig, 1917: 60ff) Gaudig did not foresee that the efforts required would focus on an *"educational ideal,"* but would instead target *"a firmly defined, inner clarity on a particular ideal of life."* (ibid: 65)

> *"Clearly and brightly ... must ... every German school have before their eyes the model of the whole German* Volk *tradition: the picture of a people bound together in a national existence by their strong consciousness of unity ...*
>
> *In powerful strokes, German 'nature' found crystal clear expression in the World War; in many respects, the new times will bring new expressions of the old nature. Through his world-historical enormous deeds and actions and his tremendous experience in these times, the German man in many respects is obligated to raise his national character* [Wesensart] *even higher."* (ibid: 64)

War was the educator of the people. That held prospectively as well, for war presented an opportunity to educate that must be utilized. The *"German people"* must acquire a future *"form of culture"* that *"is commensurate to the enormous sacrifice of blood and property in the World War."* (Gaudig, 1917a: 3) In 1916 Gaudig stated that all contemporaries of the war were obligated to *"broad, high, and difficult pedagogical thinking."* The *"grave difficulties of the times"* awakened a realization of the enormity of the future task of education and, thus, education of the people. The *"Great War"* obligated the war generation to act.

> *"Not some future generation, but the generation of the Great War must itself create the new high culture. The further we recede from war, the weaker our consciousness of war becomes, the weaker the effects of the war. For this reason, the c o n t e m p o r a r i e s - the contemporaries of the Great War must act to fulfill the great educational task; with all their drive and as soon as possible, yet before the spirit of the great time of war diminishes, yet before the great experience becomes a picture fading from memory."* (Gaudig, 1916: 4)

At the end of 1918, Gaudig advised his contemporaries in *defeat* to embrace realism, to leave off *"aestheticism,"* and to take a historical view of the events. The contemporaries were tending towards a superficial, purely emotional perception that needed to be countered by a realistic *"frame of mind"* in order to make the best of defeat. Specifically, Gaudig pointed up the *"possibility of a creative synthesis"* between a civil and a socialist social order, (Gaudig, 1918: 359) for which, once again, but negatively this time, the war as educator was responsible.

With these views, Gaudig did not in any way stand out during the First World War. He was not a spokesman, but one voice among many others. The things he had to say were heard everywhere, and there is scarce evidence that any opposing views were expressed at all. Among German educators, the building of character as a *"great task of the Fatherland"* (Colze, 1915: 12ff) was as much common property as the *topos* of war as an *"admonishment to work and to learn"* (Haumann, 1915: 271ff) or the accusation of the enemies that only they were conducting *"war pedagogy."* (Janell, 1916; 1916a) Common themes were the *"poetry of war,"* (Biese, 1915) *"German military strength"* with a view to the *"coming generation,"* (Matthias, 1915) the war as *"experience of the German national people,"* (Ziehen, 1914) and - over and over - war as the *educator of youth.* (Spranger, 1914; Kerschensteiner, 1915; Klaussmann, 1916)

The theory had consequences that reached as far as daily lesson plans at school. For instance, models were developed for patriotic essays that were used to focus German language classes on war. One of these collections of essays was prepared by Otto Karstädt, a well-known writer on pedagogy who was headmaster of a school in Nordhausen during the First World War.[47]

In 1916 he gave seventh-grade pupils the following essay as a model worthy of emulation:

> *"If I were eighteen years old, I would volunteer for service in the artillery. I would probably be sent to Erfurt or Naumburg. They would teach me to ride, shoot, fence, and drill. I would also have to learn how to shoot cannon. If I were in the calvary Fahrer, I would have to care for my saddle horse and my harnessed horse. Once I had had enough training, we would learn sharp-shooting again in the town of Ohrdruf. Then I would be clothed in the field gray uniform and go with some regiment into battle. I would act bravely and want to receive the 'Iron Cross.' If it is God's will, I would be happy to die a hero's death for the Fatherland."* (Karstädt, 1916: 25)

During the war, sayings and drawings by children were offered as entertaining readings, (*Die Kinder und der Krieg*: Floerke, 1915) school music, like the school essay, was also politicized, (Lemmermann, 1984: 308ff) and the book *German Students' War Letters* (Witkop, 1915) became such a successful literary bestseller that it appeared as late as 1933 in a *"special edition for the public."*

[47] Otto Karstädt (1876-1947) attended the Teaching Seminary in Osterburg and worked as a primary and secondary schoolteacher. He went on to study at Leipzig und Jena. From 1918 to 1929 Karstädt worked as head of a government department in the Prussian Ministry for Culture and was then appointed to the Pedagogical Academy in Hanover. Karstädt was best known for his 1919 book, Methodische Strömungen der Gegenwart, published in a total of 19 editions up to 1931.

Even elementary teaching was transformed into a military enterprise. (Göbelbecker, 1914)

Wilhelm Rein also joined this united pedagogical front, writing in 1916:

> *"This is what distinguishes the German people from the other nations: we believe in working together, supported by ideas. Today this has appeared in tangible form: The German Volk is a tremendous Work Community in which the individual takes his integral place in order to serve the whole. In this all-inclusive organization, which represents an organism, lays our strength. While our enemies may mock this as "militarism of the German type"* (Deutschtum) *we are proud of the connection between war work and peace work. In commonly shared esteem of the voluntary discipline of a great people's army, war army, and work army, we bind together tightly. This is one of the fruits of the Great War."* (Rein, 1916: 17)

Connections Continuing after the First World War

In 1914 Othmar Spann[48] sought to raise the fundamental notion of Gliedschaft (incorporate parts or members of a whole) to the level of a sociological theory, which met the approval of Peter Petersen, a reform pedagogue in Hamburg who was little known at the time. Petersen took reference to this theory again in 1924 in his Hamburg lectures on General Educational Science.[49] (Petersen, 1962:

---

[48] Othmar Spann (1878-1950) received a doctorate in socio-economics in 1903 under Friedrich Neumann and Arthur Schäffle in Vienna. His habilitation as professor of national economics was accepted in 1907 at the *Deutschen Technischen Hochschule* in Brünn, where he was associate (1909) and full (1911) professor for national economics and statistics. In 1909/1910 Spann organized the Austrian national census. He was appointed to the Political Science Faculty in Vienna in 1919, where he served as professor for philosophy and sociology up to 1939. His theory of the nation of social estates [*Ständestaat*] (*"Der wahre Staat,"* 1921) was intended as a correction to Western liberalism and to trigger the return of social theory to political romanticism, particularly to the ideas of Adam Müller. On 23 February 1929 Spann held the keynote speech at the inauguration of the newly founded *Kampfbundes für deutsche Kultur* in the presence of Adolf Hitler and Alfred Rosenberg. Previously, in 1926, Spann's theories had been strongly criticized at the Fifth German Sociological Meeting. Spann's initial political engagement was for a Christian nation of social estates and initiated the "Korneuburg Oath" of the Austrian Army published in 1933. From 1933 there was some attempt at convergence towards National Socialism, organized at the *Institut für Ständewesen*, which had been sponsored by Fritz Thyssen. However, Spann's theory of the social estates was condemned by the National Socialists. After German troops invaded Austria, Spann was arrested, and he was forbidden to teach following his release.

[49] Petersen's *Allgemeine Erziehungswissenschaft* was first published in 1924 and was based on lectures that he gave in 1921 and 1923 at the University of Hamburg. Peter Petersen (1884-1952) was headmaster of the Lichtwark School in Hamburg and had completed his

18ff) It did not trouble Petersen in 1934 to point out the psychological significance of political, including military, symbols that should be utilized pedagogically. The flag educates, because it creates emotional unity, "political enthusiasm" for the people as pointed out later. (Petersen, 1937: 173) Leadership requires sublime "ceremony." (ibid: 193)[50] And, as it was held in Petersen's circle in 1936, "the success of a bid for leadership is determined by loyal followers." (Mey, 1935: 31)

Einfügung in die Gemeinschaft [Embodiment into the Community] was as much a problem within German pedagogy after 1933 as it had been at the turn of the century, and it was then only a short step to reestablish a connection to the *war pedagogy* of 1914-1918. (Krumm, 1935) Erich Weniger,[51] along with others, attempted to do so in a work of 1938.[52] He refers to the topos of *War as Educator* and, with this, is satisfied to have found the solution to a fundamental pedagogical problem. Just as there is no *"born educator,"* there is also no *"born soldier."* But as *"comforting, basic pedagogical experience"* has shown, one can be taught to be a soldier, namely, by war: (Weniger, 1990: 208)

> *"War itself, with its severe and absolute demands, turns thousands and thousands of boys and men, as they fulfill their war obligations, into good soldiers. Without war they would have followed their many-sided natural abilities and obeyed their manifold peaceful tendencies, in commonplace fates as citizens, farmers, or workers. The majority would not have felt the calling to be soldiers even if they had served in*

habilitation for teaching philosophy at the University of Hamburg in 1920. Petersen was appointed as Wilhelm Rein's successor at the University of Jena in 1923.

50 *Führung verlangt den „Aufstieg zur Feierstunde."* (Petersen, 1937:193)

51 Erich Weniger (1894-1961) studied 1913/1914 first in Tübingen and from 1919 on in Göttingen. Weniger received a doctorate in 1921 under historian Karl Brandi with a dissertation on Rehberg and Stein. In the same year he became head of the institute of adult education in Göttingen founded by Herman Nohl (1879-1960) who was appointed as professor of philosophy of education in Göttingen 1920. In 1923 Weniger became Nohl's assistant and completed his habilitation with a paper on *The Foundations of Teaching History* in 1926. He was appointed to the Pedagogical Academy in Kiel in 1929 and in 1930 became the director of the newly founded Pedagogical Academy in Altona. When it closed, he was transferred as director of the Pedagogical Academy in Frankfurt am Main. Here Weniger was suspened (May 1933) and then dismissed from office (September 1933) as *"politically unreliable"*. Weniger, however, managed to have the dismissal reversed, and in January 1934 he was named *"secondary school teacher with civil servant status"*, retaining his former title of office and his former rank. He taught since the beginning of 1938, after being granted time off for scientific research. From 1939 on, Weniger served as captain and then major in the *Wehrmacht*, including serving on the staff of General Heinrich von Stülpnagel.

52 *Wehrmachtserziehung und Kriegerfahrung* (Berlin, 1938; excerpts in: Weniger, 1990: 202-269).

some military capacity in peacetime. But, when the extraordinary event fell over them, they survived it well and bravely, in a manner far exceeding that which could have been expected on the basis of their inherited abilities and prior tendencies. They became real *soldiers."* (ibid.)

War puts education to the test, for *"only that which has proven valuable in the face of the reality of war should find a place ... in education in peacetime"* (ibid: 209). The state of peace directly follows the state of war; the attitudes that war has engendered inform the period of peace. War has pedagogical yields that, subsequently, must be utilized.

Despite *"war's educational power,"* (ibid: 210) Weniger also makes a plea for specific training for soldiering (ibid: 215). This was the only route to achieving the *enhancement of will* [Willenssteigerung] with a view to the *"self-assertion of* Volk *and state."*[53] (ibid) Education applies to the military form of life, in which a particular *"attitude of will"* of the people is expressed. This is

> *"ultimately mental behavior* [geistiges Verhalten] *that cannot be transmitted mechanically, but can be demonstrated by way of example, or encountered in life, such that its meaning becomes readily apparent and clear"* (ibid: 217).

With this, Weniger has transposed the categories of a theory of education that was developed within the life philosophy pedagogy of the Weimar Republic.

This behavior arises from the *"spirit of the military (Geist des Soldatentums) which is taken up within the ethos of the personality, and in which the soldier feels at home as something with an objective reality to which he has moral obligations."* In the long term this behavior is *"secured"* by a *"fundamental military attitude (Soldatische Grundhaltung), a sense of the nature and essence of military priorities and decision-making, which triggers soldierly judgments when faced with any position or conflict."* This basic attitude is the target of education:

> *"Military manners and moral standards serve to secure the basic attitude, and they are the product of education and training. The fundamental military attitude and its ethos in the reality of soldierly behavior becomes revealed in military geniuses, those great soldiers who with their persons have represented the model soldier - much*

---

53 *"Ultimately ... soldiering like educating is not at all about the standard of living of particular classes based mainly on inherited predisposition and natural tendencies, but about an* attitude of will, *which is demanded by the historical necessity of the self-assertion of the people and state, and about a corresponding total* way of life." (Weniger, 1990:215)

> *more so than in their theories and instructions, which often enough have been developed with insufficient means."* (ibid: 218)

The agents responsible for this education should be the *"war generation,"* that is, participants in the First World War who can transmit the *"true contents of the war experience."*[54] (ibid: 233) For they indeed share *"despite any differences in their experiences and memories, one basic experience ... in which is contained the real and true meaning of the war for our* Volk.*"* (ibid: 234-235)

> *"This crucial experience, which in retrospect becomes ever clearer, was that the average man was exposed to the extraordinary (and that, over the long run), that he grew into the community and into camaraderie, that he had to live in the long run outside all normal regards of life and to the very roots of his existence cut off from them and, at the same time, stood directly in the path of death, such as otherwise is experienced only by extraordinary men under extraordinary fates."* (ibid: 235)

The heroic in the everyday of war had the consequence that *"a completely new type of calm and courageous life"* could develop, which allowed *"life at home* [Heimat] *to be determined by real life on the battlefield."* Being *"on the battlefield"* was a special type of community, namely, *"the people at arms."* (ibid)

> *"All of the great qualities, duty, loyalty, camaraderie, and all human experiences took reference from this incredible ever-present state of being. All great paroles became groundless. Even to speak of sacrifice, as countless and enormous the sacrifices were, misses the essential meaning. It is not about sacrifice for others, but something much more immediate."* (ibid.)

What Weniger means here is the bond with the people. In sum, he wrote: *"The insoluble unity of military life (Soldatisches Dasein) and the whole life of the* Volk *is ... one of the crucial experiences of the World War."* (ibid: 236)

In a 1936 *"Honorary Book for German Schoolteachers"* [*Ehrenbuch deutscher Volksschullehrer*] this thesis was to be underpinned from the other side, the *Heimat* [homeland], by using specific materials from schools and teachers of the time of the First World War. (Führen 1936) The book contains a preface by the director of the National Socialist Teachers' Association, Fritz Wächtler.[55] He writes:

---

[54] Weniger volunteered for the artillery in 1914 and returned from the war as a second lieutenant.

[55] Fritz Wächtler (1891-1945) was Minister for Public Education in Thüringen for the National Socialist German Workers' Party [*NSDAP*] in 1932/1933 and became head of the

"During the great struggle of 1914-1918, the hour of experiencing the community of the people in the camaraderie of the grave, the hour of birth of National Socialist idea in the heart of the unknown Private Adolf Hitler, every rank of our people fulfilled its duty. And here we may especially name the German educators, when we speak of the heroism of the occupations in the Great War. Many thousands of us stood for the four years of the war unshakeable, as men and as officers on all fronts, to fight, to bleed, and to die." (ibid: VII)

But not only at the Front, but also and particularly at home, those men served *"to whom it was denied, with weapon in hand to show their mettle,"* the people's mettle. That was what the book was to be about. And indeed, the book speaks of *"war songs for children,"* (ibid: 52ff) *"taking up collections for our men in field gray"* (ibid: Plate 11), pupils who went to the streets as *"canvassers for the war loan,"* (ibid: Plate 13) *"victory celebrations"* at school, (ibid: 107ff) paramilitary exercises in *"summertime,"* (ibid: 128ff) and *"sacrifice days"* at school for collection activities. (ibid: Plate 23)

A special chapter of the book is devoted to the topic of *"war poems by German teachers."* (ibid: 363-385) Unlike the aftermath of World War II, the period after 1918 saw the emerging of an immense body of literature in affirmation of the Fatherland and war. Weniger's concept of *"Grunderlebnis"* [Key Experience] is a idealization that is in no way based on real experience of the First World War.[56] But it refers to a broad body of literature that in the honorary book of 1936 is recalling to the reader teachers' positive service to the community during the war.[57] The Silesian teacher Hermann Stehr[58] was said to have provided, with his poem *Der Krieg bricht los* [Outbreak of War], *"one of the best representations of the onset of the war,"* which should be viewed as

---

NSDAP administrative district of Bavarian Ostmark in Bayreuth. Wächtler succeeded Hans Schemm as director of the National Socialist Teachers' Association in 1935.

56 See Ullrich & Ziemann (1994) documented daily existence on the Front in the First World War.

57 This includes works by such authors as *"Bauern- und Volksschriftsteller"* (farmers' and people's author) Gustav Schröer (1876-1949), Silesian nature poet Wilhelm Müller-Rüdersdorf (1889-1945), Mecklenburg author Hans Franck (1879-1964), Low German poet Wilhelm Scharrelmann (1875-1950), and also the originally Social Democratic and later National Socialist poet Herrmann Claudius (1878-1980). All of them were all public school teachers.

58 Hermann Stehr (1864-1940) attended the Teachers' Seminary in Silesian Habelschwerdt and worked as a schoolteacher from 1887. From 1915 on he turned to writing. Stehr's work encompasses mainly natural mystical novels set in the Silesian landscape. Stehr worked for the National Socialist (NS) journal *Das Innere Reich* for all eleven years of its existence (1934/35 to 1945) and in 1933 was seen as a true *"blood and soil"* poet by NS literary critics. Stehr published an autobiography, *Mein Leben,* in 1934.

pedagogical. Thus, also the war poem educates, here by presenting the scene immediately prior to the *"outbreak"* of war:

"Der Krieg schlief fest in einem Furchengraben,
Zehn Schuh tief in der Erde, und es bog
Sich wiegend über ihm die goldne Gabe
Des Korns im Sommerwind; rotflackernd flog
Der Mohn; das Bienlein trug voll Fleiss zur Wabe;
Die Wachtel schlug, der Schnitter dengelnd zog
Die Sense auf dem Amboss hin und her,
Und Deutschlands Herz war fried- und segensschwer"
(ibid: 371-372).

[War slept fast in the furrows of the field,
Ten feet deep in the earth, and bending over it
The golden gift of grain waving
In the summer wind; fluttering red, the poppies
Flew; the hard working little bee bore nectar to the honeycomb;
The quails' beat their wings, the hay-cutter honed his scythe back and forth across the anvil,
And Germany's heart was filled with peace and blessings]

In the Third Reich it was customary to dedicate poems to the *"Führer"* (Bühner, 1939), and during the first years of the war, *"poems of soldiers"* were once again used to educate the people. (Ziesel, 1940) At the same time, the basic *"image of war in German thinking"* was made the subject of discussion. (Faust, 1941) This makes it appear fitting that in 1942, one of the contributions to a festschrift in honor of one of the most famous German thinkers in education, namely Eduard Spranger,[59] addressed the topic of the *"philosophy of total war."* [60] (Wenke, 1942) Wenke's contribution attempts to make fruitful use of Carl

---

[59] Eduard Spranger (1882-1963) completed his qualifications as professor of pedagogy and philosophy in 1909 at the University of Berlin. In 1911 he was called to the University of Leipzig, returning to Berlin in 1920. He accepted an appointment at the University of Tübingen in 1946, where he taught up to 1954. During the Weimar Republic, Spranger was one of the founders of philosophical pedagogy in Germany, and he advanced to become one of the most eminent pedagogues in the new Federal Republic of Germany.

[60] The author, Hans Wenke (1903-1971), was a pupil of Sprangers and completed his habilitation in Frankfurt in 1938 with a paper on the *"National Socialist New Order of the German School."* (Wenke, 1944:47ff) In 1940, Wenke was appointed Professor of Pedagogy and Philosophy at the University of Erlangen and in 1947 at the University of Hamburg, and in 1949 Full Professor of Philosophy and Pedagogy at the University of Tübingen. Here he served as president of the university in 1953/1954. Wenke was cultural senator of the City of Hamburg from 1954 to 1957. In 1958 he was given a chair at the University of Hamburg. His essay on re-education, Education in Western Germany: A Postwar Survey (1953), was added to the Library of Congress. In 1968, Wenke left his chair, following severe conflicts with the Hamburg student movement.

Schmitt's theory of the *enemy* for the *"great task of political education."* This was a task that transcended far beyond school lessons, and related to the people as a whole, being in this respect *total*: (ibid: 288)

> *"Germany sought realization of the total war taking very different routes than England did. It understood and realized the totality, in accordance with a deeper understanding in its philosophy and worldview, as the true unity in the multitude; that is, it placed the total war upon the wide and broad fundament of its own people, in all its parts and members. This total war rules our times."* (ibid: 289)

Citizens in Uniform and the Setting-Up of Taboos

In the early 1950s, the slogan *Bürger in Uniform* [Citizen in Uniform] was discussed in the Federal Republic of Germany. Erich Weniger, who was to become one of the most influential educational theorists after the war, contributed to the discussion, advising conservative politicians on the *"rearmament,"* as it was called, as the entire German armed forces had been disarmed in 1945.[61] Weniger, who was appointed as Herman Nohl's successor on the leading chair of philosophy of education in Göttingen in 1949, was the main speaker at the *"Weinheimer Discussions"* in 1951 und 1952, at which the new concept of *Bürger und Landesverteidigung* [Citizen and National Defense] (1952) was presented and discussed.[62]

Erich Weniger and Werner Picht,[63] adult education specialist and journalist, discussed the idea of the Citizen in Uniform in 1953 in the Göttingen journal, *Die Sammlung*.[64] The previous year, Picht had published a critical work on the *"war nature of the Germans"* and took it as a foregone conclusion that an ethos of the civilian who, in the absence of an authoritarian state, could be entrusted to wear a uniform and bear arms was non-existent in Germany. The societal prerequisites for a democratic army did not exist, and the concept of the Civilian in Uniform lacked a basis. Moreover, Picht wrote that any political integration of the military stood in contradiction to its *"autonomy,"* it having laws unto itself.

---

[61] In 1955 Weniger became a member of Personnel Evaluation Committee of the Bundeswehr, which selected among candidates for Senior Officer Positions. Weniger also became a member of the Bundeswehr Advisory Council for Internal Management in 1958.

[62] Weniger spoke on *"Social Problems: A Contribution towards European Defense."*

[63] Werner Picht (1887-1965) was advisor on adult education centers and thus adult education to the Prussian Ministry of Culture from 1919 to 1924. From 1927 to 1933 he was director of the higher education department of the League of Nations Institute of Intellectual Cooperation in Paris. Picht became the director of the German Academic Foreign Exchange Students Service in 1935. In 1937 he published a paper on soldiers at the Front.

[64] Weniger (1953, 1953a), Picht (1953).

Defense was a question of an *"able-weaponed team"* and not integration into a democracy. (Picht, 1953)

Weniger justified the new task of the armed forces with the idea of *"self-defense of the civilian"* in the framework of European integration and under the prerequisite of a democratic identity and constitution, which every civilian has the responsibility to uphold. (Weniger, 1990: 300) The mistake made by the *"old army"* had been to *"completely remove the soldier during military service from the thought and political contexts of his civilian life,"* whereas it was now time to seek precisely that binding of the armed forces to broader society. Military service was no longer service to the *"Fatherland,"* which in the past has been abused to serve *"basic conservative views,"* (ibid: 305) but service to democracy.

This formulated a policy priority, according to which the military could no longer be *"state within a state,"* nor could it be obligated by oath to a *Führer.* (ibid: 306) Instead, civilians must wear their uniforms for their system of government. This concept also provided a basis for universal male conscription, as a further clamp holding together the army and society:

> *"The new armed forces of Germany must ... express the spirit and ideas that stand at the basis of our political existence, must themselves be the most vital embodiment of this spirit in military form. They may not oppose the whole or parts of the people, who represent their views and goals in the framework of the rules of a democracy."* (ibid: 306)

In one of his last works of 1960, Erich Weniger expressed his opinion regarding the military tradition in democracy [Soldatische Tradition in der Demokratie].[65] His first point is that none of the earlier forms of the army of the German Reich – that is, the Imperial Army, the Reichswehr (1921-1935), and Hitler's Wehrmacht – should serve as a model for the Federal Armed Forces [Bundeswehr]. (ibid: 349) If any links to tradition could be forged, they would be to the "Prussian Reformers, who prepared the way for and made possible the Wars of Liberation"[66] (ibid: 353) Other role models were the "martyrs" of 20 July 1944, the attempted assassination of Hitler, whereby Weniger at the same time reminds us of the "high war performance" and the "great achievements, even in the face of superior forces," in "the last war." (ibid: 354) It was important that "living memory" lead to *"critical self-reflection, which, through clarifying the experience, allows us to make a new beginning."* (ibid: 354-55)

In Germany, in the face of the past, this appears to be difficult if not impossible. In the field of education, *"military education"* has become taboo, which seems

---

[65] First published in: *Die neue Gesellschaft*, 7. Jahrgang, Heft 3 (1960), 196-203.
[66] Wars against Napoleon 1813 to 1815.

understandable if we recall the martialistic strivings that characterized the theory and practice of military education from the German Empire up to military education in the German Democratic Republic (GDR). That form of education is not compatible with a democratic constitution, or in other words, with open forms of society. It is not a coincidence that paramilitary education in the GDR-schools[67] did not find public acceptance, since it stood in contradiction to socialist traditions and corresponded precisely with the kind of military education that had been built up in Germany from the turn of the century.

Up to 1945, and in the GDR up to 1989, military education focused on conformity in thought, coupled with a strict distinction between friend and enemy, and took reference to national territory. This was based on the threat and conquest scenarios of the first half of the twentieth century that, with the end of the Second World War, became increasingly obsolete. The *Bundeswehr* was still charged with defending the freedom of the nation-state under the framework of the alliances with no other scenarios in sight. (ibid: 339) National defense was always conceived as defense and deterrence against a compact enemy that threatened the whole territory of the state. Lifting the borders of war after the Second World War meant lifting the borders of the battleground, whereas the outcome of the two world wars had been decided through subjugating and occupying enemy territory.

That is largely, although not completely, a thing of the past. *"Wars"* will not be conducted in future by armies facing each other from behind permanent lines. Military conflicts now increasingly resemble highly armed, but limited, strikes against enemies who resemble *"soldiers"* less and less. What is also new is that while certain operations and battles have a start and a definitive end, many *"wars"* do not. They continue on, sometimes as unresolved conflicts over decades, or as regular eruptions of violence, or as a regional strategy to solve conflicts of power. Wars between nations will likely recede; outcomes - victory or defeat - will become rarer, since in guerilla-type conflicts it is only with difficulty that the quality of victory or defeat can be determined. In asymmetric war conflicts, it is not simply the stronger party that wins.

In addition, the setting of *"military education"* in states conducting war has shifted to the media. Education as presentation in the media no longer attempts to teach military *"attitudes,"* but instead aims to gain control of public opinion, particularly control over the power of the images of war, whatever the character

[67] Starting from school year 1978/1979, *"Military Defense"* was a required school subject for the ninth and tenth grades at the unified comprehensive schools. The subject matter focused on a clearly defined enemy, which was to be internalized by the pupils. This *"education of youth to take up weapons"* corresponded exactly with the traditional militaristic topos of *"Erziehung zur Wehrhaftigkeit"* [education towards the ability to defend oneself militarily].

of the war may be. Along the same lines, in contrast to what Eduard Spranger had thought in 1914, young people do not have to be won over for war. Instead, shifting percentages of public approval or disapproval emerge, which requires the existence of responsible and discerning citizens, who must be informed of the risks and repercussions of any specific *"military intervention,"* as they are now called. With this, military education is a question of public education, and it does not involve following those goals that, at least in German pedagogy, targeted the people as a whole as their object.

For the All-Volunteer Force, military training is a question of professional training and no longer the task to *"educating the people."* Military education as a general educational issue does not have the militarization of society or school as a prerequisite, but instead takes its reference from the problems of international security, which have changed in nature radically and will continue to change. An understanding of future conflicts and associated military options are a part of political education, which must address and adapt to a globalized world. At the same time, democracy is the basis of all value decisions. The problem in German pedagogy lies in the fact that this connection was recognized only as late as 1945 and developed only hesitantly.

Prior to and especially during the course of the First World War, war education was *"totalized,"* in the sense that it was in fact held that a war orientation must be induced in all children and the people as a whole. This is not only a German phenomenon, but a European one (compare, e. g. Audoin-Rouzeau, 1993 on France) that found similar forms of expression in literature and education everywhere that have no place in a civil society. The German peculiarity stems from the development of a special language of *"experience"* [*Erlebnis*] and the after-the-fact idealization of heroism, along with dramas of treason and totally unheroic self-pity. Nowhere else do we find the idealized *"Front experience"* or the myth of the *"Dolchstoss"* [stab in the back] of the *"Heimatfront,"* and nowhere has there been such intensive thinking about war education as in Germany. It makes sense, therefore, that people are allergic to the topic.

These dictates of history, however, do not justify the tabooing of military problems in the field of education today. On the contrary, the foreseeable, new security threat scenarios; the strategies of media influence; and, not least, the need for citizens to be informed, all demand that security issues become educational issues. This requires that German military education belong to the past, while at the same time, we can no longer allow military issues to be ignored. They are part and parcel of society, no one can avoid them, and it is necessary to deal with them as rationally as possible. In Germany, this requires a historical break with the organization and mentality of *"military education."*

War is no longer simply the extension of politics by other means. (Clausewitz, 1980: 34) Instead, every war, no matter what form it will take, must be supported by a complex cluster of interests that are articulated politically, without already dictating the means. Indeed, peace activities can also be effective means of implementing policy; for situations requiring policy decisions there is a whole range of options available, including military options, that may or may not be implemented, depending on their expected effectiveness. Military means are never the only means, and they are also not the only means of maintaining power. Demonizations that maintain the contrary do not contribute usefully towards rational evaluation of security or threat scenarios.

In other words, with regard to military education, German education should absolve itself of the double spell of its ill-fated tradition and the resulting taboos, whereby it should be added that tabooing was always connected with subliminally effective theories like the *political* theory based on Carl Schmitt. Precisely because politics *"ultimately"* seemed to be defined by authoritative decisions as to *"friend"* or *"foe,"* not by attenuation of conflicts, only radical solutions appeared to be possible, namely *"principal"* belligerence or pacifism, whereby German experiences with the history of military education clearly indicated the choice.

Today, however, the educational realities encompass the training of children as soldiers (Child Soldiers Report 2001), military-religious oppression, such as of the Taliban children, children growing up in war zones (e.g. Garbarino; Kostelny; Dubrow, 1991), the misery of child refugees (Kibreab, 1996), or the untapped potential of adolescents affected by armed conflict (Lowicki, 2000). It is not by coincidence that the World Bank issued a report in 2000 on the relations among Violence, Democracy and Education (Salmi, 2000). Placing a general taboo on these connections does not help at all.

What is needed are educational solutions that start out from the assumption that education is a means to counteract de-civilization. For German education, this would truly be a fresh start.

References

Biese, A. (1915): Poesie des Krieges. Berlin, Grote.

Bühner, K. (eds.) (1939): Dem Führer. Gedichte für Adolf Hitler. Stuttgart; Berlin, Georg Trunckenmüller Verlag.

Clausewitz von, C. (1980): Vom Kriege. Hinterlassenes Werk. Frankfurt; Berlin; Wien, erste Ausgabe 1832-1834, Ullstein.

Colze, L. (eds.) (1915): Universität und Schulen im Kriege. Berlin; Leipzig; Wien, Arthur Collignon Verlag (= Deutsche Kraft. Kriegskultur und Heimatarbeit 1914/1915, hrsg. v. L. Colze. Heft 15).

Dahn, F. (1892): Moltke als Erzieher. Leipzig.

Floerke, H. (eds.) (1915): Die Kinder und der Krieg. Aussprüche, Taten, Opfer und Bilder. Mit 21 Bildbeigaben und Kinderzeichnungen. München, Georg Müller.

Faust, A. (eds.) (1941): Das Bild des Kriegs im pädagogischen Denken. Bd. I, Stuttgart; Berlin, W. Kohlhammer.

Führen, F. (eds.) (1936): Lehrer im Krieg. Ein Ehrenbuch deutscher Volksschullehrer. Die Kriegsarbeit deutscher Lehrer in der Heimat. Mit hervorragender Unterstützung der Gliederungen des NSLB, der Lehrerbüchereien, staatlicher und parteiamtlicher Dienststellen und vieler Volksschullehrer, sowie im Einvernehmen mit der Reichsamtsleitung des NSLB bearbeitet. Berlin, Kolk.

Gaudig, H. (1916): Wir Zeitgenossen. Einige pädagogische Neujahrsgedanken. Zeitschrift für pädagogische Psychologie und experimentelle Pädagogik 17, 1-5.

Gaudig, H. (1917): Deutsches Volk - Deutsche Schule! Wege zur nationalen Einheit. Leipzig, Quelle & Meyer.

Gaudig, H. (1917a): Pädagogisches Denken und nationales Leben. Einige Neujahrsgedanken. Zeitschrift für pädagogische Psychologie und experimentelle Pädagogik 18, 1-4.

Gaudig, H. (1918): Die Denkverfassung der deutschen Seele in der Zeit der grossen Krisis. Zeitschrift für pädagogische Psychologie und experimentelle Pädagogik 19, 353-360.

Haumann, G. (1915): Der Krieg und die Jugendlichen. In: Staatsbürgerliche Belehrungen in der Kriegszeit. Herausgegeben für Fach- und Fortbildungsschulen vom Kgl-Preuss. Landesgewerbeamt. Berlin, Carl Heymanns Verlag, 265-274.

Janell, W. (1916): Kriegspädagogik. Berichte und Vorschläge in Verbindung mit Dr. Walther von Hauff, Georg E. Kik, Dr. Otto Nothdurft. Leipzig, Akademische Verlagsgesellschaft.

Janell, W. (1916a): Kriegspädagogik unserer Feinde. Der Panther. Deutsche Monatsschrift für Politik und Volkstum, 4. Jg., H. 12, Dezember 1916a, 1566-1579.

Jünger, E. (1981): In Stahlgewittern. 27. Aufl., Stuttgart, erste Ausg. 1920

Kerschensteiner, G. (1915): Der Krieg als der grosse Volkserzieher. Zeitung der oberösterreichischen Land-Lehrervereine Jg. 1915, 7ff.

Klaussmann, A.O. (1916): Der Krieg als Erzieher. Berlin, Siegismund.

Lamszus, W. (1913): Das Menschenschlachthaus. Bilder vom kommenden Krieg. Hamburg; Berlin, Alfred Jansen.

Matthias, A. (1915): Deutsche Wehrkraft und das kommende Geschlecht. Leipzig.

Meinhold, E. (1908): Deutsche Rassepolitik und die Erziehung zu nationalem Ehrgefühl. München o.J.

Nohl, H. (1960): Berthold Otto. Die Sammlung 15, 215-216.

Petersen, P. (1934): Psychologische Bedeutung der politischen Symbole. Zeitschrift für Jugendkunde IV, 1, 10-14.

Petersen, P. (1937): Führungslehre des Unterrichts. Langensalza; Berlin; Leipzig, Verlag von Julius Beltz.

Petersen, P. (1962): Allgemeine Erziehungswissenschaft. Zweite, unveränd. Aufl., Berlin, Walter de Gruyter&Co. (Erste Aufl. 1924).

Picht, W. (1937): Der Frontsoldat. Stuttgart.

Picht, W. (1952): Vom Wesen des Krieges und vom Kriegswesen der Deutschen. Stuttgart.

Picht, W. (1953): Bürger in Uniform? Die Sammlung 8, 392-396.

Rein, W. (eds.) (1897): Encyklopädisches Handbuch der Pädagogik. Vierter Band, Kadetten-Anstalten – Myopie, Langensalza, Hermann Beyer&Söhne.

Scheler, M. (1917): Der Genius des Krieges und der Deutsche Krieg. 3., neu durchges. Aufl., Leipzig, Verlag der Weissen Bücher (erste Aufl.1914).

Schmidt-Gibichenfels, O. (1912): Der Krieg als Kulturfaktor als Schöpfer und Erhalter der Staaten. Politisch-Anthropologische Monatsschrift, November/Dezember 1912, 2. Aufl. Berlin-Steglitz, Verlag "Kraft und Schönheit" (um 1916).

Sombart, W. (1934): Deutscher Sozialismus. Leipzig.

Spann, O. (1914): Gesellschaftslehre. Leipzig.

Spranger, E. (1914): An die Jugend. Der Säemann Jg. V, H. 9-11 (September/November 1914), 386-393.

Velmede., A.F. (eds.) (1937): Kriegsdichter erzählen. München, Langen-Müller.

Wagner, A. (1914): Gegen England. Berlin.

Weniger, E. (1952): Social Problems: A Contribution towards European Defense. In: Bürger und Landesverteidigung, 11-31.

Weniger, E. (1953): Bürger in Uniform. Die Sammlung 8, 57-65.

Weniger, E. (1953a): Bürger in Waffen. Die Sammlung 8, 396-399.

Weniger, E. (1990): Lehrerbildung, Sozialpädagogik, Militärpädagogik. Politik, Gesellschaf, Erziehung in der geisteswissenschaftlichen Pädagogik. H. Gassen (eds.): Weinheim; Basel, Beltz Verlag.

Wenke. H. (1942): Zur Philosphie des totalen Krieges. In: H. Wenke (eds.): Geistige Gestalten und Probleme. Eduard Spranger zum 60. Geburtstag, Leipzig, Quelle & Meyer, 267-289.

Willmann, O. (1914): Lehrstand als Wehrstand. Zeitschrift für christliche Erziehungswissenschaft Jg. 1914, 603ff.

Witkop, Ph. (eds.) (1915): Kriegsbriefe deutscher Studenten. Gotha, Verlag F.A. Perthes.

Witkop, Ph. (ed.) (2002): German Students' War Letters. Transl. by A.F. Wedd; ed. by J. Winter, Philadelphia PA, University of Pennsylvania Press.

Ziehen, Th. (1916): Die Psychologie grosser Heerführer. Leipzig.

Ziesel, K. (eds.) (1940): Krieg und Dichtung. Soldaten werden Dichter - Dichter werden Soldaten. Wien; Leipzig, Adolf Luser.

David L. PERRY, U.S. Army War College, Carlisle

HOW ETHICS IS TAUGHT AT THE U.S. ARMY WAR COLLEGE

This essay will investigate the nature and scope of ethics, the value of studying ethics, some ways in which formal ethics education is incorporated into the curriculum of the U.S. Army War College (USAWC) where I teach, and some strengths and weaknesses of our ethics pedagogy.

The Nature and Scope of Ethics

Ethics and morality - along with law and etiquette - are essentially *normative*, that is, they prescribe human behaviors as obligatory (must do), prohibited (must not do), or permissible (may do). By contrast, many disciplines like the natural sciences, psychology, history and economics are empirical, meaning that they attempt to describe, explain or predict events or motives or actions. In general, empirical disciplines deal with facts and probabilities, while normative disciplines promote or assess social values or state obligations. Empirical disciplines study what exists, what happened, or what tends to happen under certain conditions; their claims can at least in theory be tested using controlled scientific methods or in light of the best available evidence. Moral principles state how human beings ought to treat one another; moral claims cannot be proved or disproved by empirical means alone. Ethical arguments often rely on empirical assumptions, though. And moral beliefs can be surveyed empirically: that is what many sociologists and anthropologists do.

The words *"ethics"* and *"morality"* have Greek and Latin origins, respectively. Traditionally they referred to customary values and rules of conduct (meanings still present in our terms *"cultural ethos"* and *"social mores"*), as well as insights about what counts as human excellence and flourishing. Ethics and morality sometimes have different connotations in everyday conversation. But we often use those words interchangeably as well: e.g., we can usually call a moral issue an ethical issue with no change in meaning. Ethics also refers to **moral philosophy**, i.e., a discipline of critical analysis of the meaning and justification of moral beliefs.

There is considerable overlap between ethics and law, and ethics and etiquette. Much of the law embodies ethical principles: respect for basic rights to life and property, and the right of citizens to participate in political life. It is usually unethical to break the law. A lie can rise to the level of criminal fraud, and a broken promise can be *"actionable"* as a breach of contract. A breach of etiquette can also be unethical if it is done intentionally to offend someone simply for one's own amusement.

Ethics goes beyond etiquette, though, to include matters that nearly every human society considers significant: actions such as lying, breaking a promise or killing someone are almost always more serious than social *faux pas*. Ethics labels some actions as ideal or admirable, but not strictly mandatory, thus going beyond law and etiquette. Ethics also has to do with human character and motivation, which in many cases are irrelevant to etiquette and law. And law and etiquette can sometimes be criticized on moral grounds: consider U.S. laws and customs that historically treated African Americans and women as less than full citizens (to put it mildly).

The Value of **Studying** *Ethics*

I tend to assume that most adults have developed what might be called the *"standard equipment"* of ethics: sympathy toward the suffering of others, a sense of fairness and injustice, and habits of telling the truth and keeping promises even when doing so is not personally convenient; in short, a conscience. In other words, I tend to assume that none of my adult students is a coldly scheming manipulator or a psychopath. My confidence in the students at USAWC more particularly is even greater than it would be in a typical undergraduate class, given that most of the former have been trained according to high standards of professional military conduct, which includes commitments to certain ethical principles. But what then is the point of requiring senior military and civilian leaders to study ethics, if their *"standard equipment"* is already in place?

Well, one reason is that people do not always take the time to reflect on their moral beliefs and ethical decisions, and as a result they sometimes make costly mistakes that they might otherwise have avoided. Also, when a moral dilemma arises where two important ethical principles point us in different directions, our consciences may sometimes be unable to prescribe a clear resolution to the problem. Occasionally, especially in the uses of military force, we even face tragic decisions where each of the available options will result in serious harm, and we must therefore choose the lesser of evils. And people of integrity sometimes disagree fundamentally about what's right and wrong, so it won't do at that point simply to encourage them to obey their individual consciences. We need wisdom and critical thinking as well as a good conscience. We need to find some way to judge which moral views are most sensible. Moral philosophy can be helpful in these respects.[68]

---

68 For excellent introductions to this subject, read William Frankena, *Ethics*, 2nd edition, and James Rachels, *Elements of Moral Philosophy*, 4th edition.

Ever since the ancient Greek philosopher Socrates, ethicists have tried to construct a unified vision of the moral life that would include a rational means of testing the adequacy of moral claims and resolving conflicts among important moral values. Moral philosophy often proceeds by proposing an ethical principle or identifying one that is implied by a particular action or policy, testing that principle to see if it is vulnerable to counterexamples, and if so, revising the principle or proposing a new one that is less vulnerable to objection. Moral philosophy can't guarantee infallible decisions, but it ought to reduce the likelihood of making bad choices due to lack of careful reflection. That is why it is an important component of education at USAWC.

The U.S. Army War College

U.S. Army officers are exposed to formal education in ethics at many different points in their careers, including their basic and advanced officer courses, the Command and General Staff College which typically enrolls captains and majors, and USAWC which is oriented to lieutenant colonels and colonels with about 20 years of experience.[69] Many of our graduates (like their counterparts at the National War College and other senior-level schools) go on to serve in the Pentagon, and some eventually become general officers. Although most of our students are U.S. Army officers, we also have a number of personnel from the Navy, Air Force, Marines, federal civilian employees, and many military officers from foreign countries. All of our students have been chosen from a highly competitive pool of candidates. I consider it a great privilege to be able to teach - and learn from - such students.

I have been working in my present position as Professor of Ethics at USAWC since June 2003. Before then I spent ten years at undergraduate universities teaching ethics in warfare, medicine, and business. Because the USAWC is my first exposure to any military organization from the inside, though, my learning curve during the past year has been steep, as you can imagine.

My formal title should not be inferred to mean that I am the sole "provider" of ethics education at USAWC. Ethical issues and concerns are raised and addressed nearly every day by myriad faculty and administrators, as well as the students themselves. Our students come to us having incorporated ethics into their training and supervision of subordinates over a period of many years, sometimes forged in the heat of combat against enemies who do not always obey the laws of war. And every member of the USAWC community *"teaches"* ethics in the example they set in their daily lives, in treating their colleagues,

[69] For a good overview of ethics education at various levels of the Army, consult John Brinsfield, *"Army Values and Ethics: A Search for Consistency and Relevance"*, *Parameters*, Autumn 1998.

spouses and children with respect and compassion, in volunteering their spare time in the community, and in their dedication to protect and defend their country even if they must sacrifice their own lives to do so.

Our academic year is structured very differently from other U.S. military schools. USAWC students are assigned when they arrive in late July to a seminar group of about 17, led by three faculty from different departments, with whom they work in a series of four required *"core"* courses that continue through January. So Term I is considerably longer than the other two terms. Only in February are the students first allowed to take elective classes. In the spring they work intensively on a Strategic Crisis Exercise, where they play roles of senior military and civilian officials who are forced to deal with a challenging series of fictional but realistic crises, from major regional wars to terrorist attacks to natural disasters. Most students will complete a master's degree in strategic studies as a result of their academic work here.

Core Courses

During the first core course in Term I, **Strategic Leadership**, one three-hour lesson is devoted to ethics. This year that lesson included a series of articles by Dr. Martin Cook, my predecessor as professor of ethics: *"Moral Reasoning as a Strategic Leader Competency;" "Moral Foundations of Military Service;"* and *"Religion and Morality: Exploring the Connections."* Some of our faculty supplemented or replaced Cook's articles in that lesson with three of mine: *"An Introduction to Ethics;""Ethics and Warfare in Comparative Religious Perspective;"* and *"Just-War Principles and the War in Iraq."* In 2003 we also inaugurated an annual lecture series on ethics, created to honor our former Chaplain (COL) Sonny Moore and his wife Martha. This lecture occurs in conjunction with the Course 1 ethics lesson.

In previous years a separate lesson in Course 1 dealt with just-war principles. Dr. Cook had urged that that lesson be split into two, one on *jus ad bellum* criteria and the other on *jus in bello* rules, each one to be covered in two later core courses. That was a sensible idea, though it also risked dilution or even distortion of the materials once those lessons were assigned to faculty in other departments who may lack expertise in the just-war tradition.

For the Course One ethics lesson in 2004-2005, I intend to switch to Anthony Hartle's fine book, *Moral Issues in Military Decision Making*. Hartle's book (which was reissued in a fully revised form in May 2004) will provide us with a much more comprehensive and coherent approach than we had this year with our disparate collection of articles. Hartle carefully explores the concept of a profession, why the military qualifies as a profession, what the core values of the

American military profession are, and how those values relate to broader values in American society as well as the international laws of war.

This year's version of Course 1 also addressed ethics in other lessons on leadership traits and military professionalism, drawing in part on an excellent document published by the Center for Army Leadership at the Army Command and General Staff College: Field Manual 22-100, *Army Leadership*. Our Course One curriculum focuses primarily on its seventh chapter on Strategic Leadership, but our students also benefit from reading the rest of the book as well, even if they have previously encountered it in other Army institutions.

To illustrate, chapter two of FM 22-100 explains seven key **Army values** that are said to be essential to the character of ethical leaders at all levels: **loyalty, duty, respect, selfless service, honor, integrity, and personal courage**. The manual also makes effective use of stories from American military history at many different points to illustrate how those Army values have been exhibited by real soldiers and officers. None of those values will be new to 40-something Army officers who comprise most of our student body, but I think it's a good idea for us to be reinforcing them at that level.

Here are a few excerpts from FM 22-100 which are particularly interesting:

> *"Respect for the individual forms the basis for the rule of law, the very essence of what makes America. In the Army, respect means recognizing and appreciating the inherent dignity and worth of all people.... Respect is also an essential component for the development of disciplined, cohesive, and effective warfighting teams. In the deadly confusion of combat, soldiers often overcome incredible odds to accomplish the mission and protect the lives of their comrades. This spirit of selfless service and duty is built on a soldier's personal trust and regard for fellow soldiers....*
>
> *Personal Courage ... Physical courage means overcoming fears of bodily harm and doing your duty. It's the bravery that allows a soldier to take risks in combat in spite of the fear of wounds or death ... In contrast, moral courage is the willingness to stand firm on your values, principles, and convictions - even when threatened ... Leaders who take responsibility for their decisions and actions, even when things go wrong, display moral courage."* (FM 22-100: 2-10)

Although the concept of *"physical"* courage is puzzling, the distinction that the authors intend to convey between courage in the morally neutral sense and moral courage is sensible. The writers go on to illustrate moral courage through the story of helicopter pilot Hugh Thompson who risked his life to stop his fellow American soldiers from massacring civilians at My Lai in 1968.

The book also notes that there are times when disobedience will be the right choice:

> *"[A] good leader executes the boss's decision with energy and enthusiasm. The only exception to this principle is your duty to disobey illegal orders. This isn't a privilege you can conveniently claim, but a duty you must perform. If you think an order is illegal, first be sure that you understand both the details of the order and its original intent. Seek clarification from the person who gave the order. This takes moral courage, but the question will be straightforward: Did you really mean for me to ... steal the part ... submit a false report ... shoot the prisoners? If the question is complex or time permits, consult legal counsel. However, if you must decide immediately - as may happen in the heat of combat - make the best judgment possible based on Army values, your experience, and your previous study and reflection. You take a risk when you disobey an illegal order. It may be the most difficult decision you'll ever make, but that's what leaders do ..."* (ibid.: 2-23)

There are a couple of problems with the manual's list of Army values, though. One is that there seems to be too much overlap between duty, honor, and integrity to warrant listing them separately. Another problem is that the principle of compassion, e.g., for wounded enemy soldiers and civilians, should really be listed explicitly rather than simply assumed to be covered by the stated values. Both of these problems may be rooted in an apparent desire of the writers to create a list of values that could be called to mind easily through the mnemonic device, LDRSHIP. But they tend to reflect something less than ideal critical thinking about ethics.

Also, in a later section of the text the importance of good ethical decision-making is stressed, but the subject is not given the depth it deserves. The basic method recommended is: *"Define the problem; know the relevant rules; develop and evaluate courses of action; [and] choose the course of action that best represents Army values."* (4-8) That is not terribly helpful, in my view (though no method of ethical decision-making could be foolproof or comprehensive), so I have felt the need with my own seminar students at least to supplement that sketchy list with a more thorough treatment of ethical theory.

I have also sought to strengthen future AWC's lessons on Critical Thinking, adding significantly more readings and exercises on evaluating arguments and recognizing logical fallacies, which I assume will also enhance our students' skills in ethical analysis. This will also address a weakness in FM 22-100, *Army Leadership*, which deals too superficially with critical thinking.

The second core course at USAWC, called **War, National Security Policy, and Strategy**, contains a three-hour lesson on the use of force combining ethics, just war theory, and international law. In 2003 that lesson explored the following readings: Martin Cook, *"Ethical Issues in War: An Overview"*; Kenneth Anderson, *"Who Owns the Rules of War?"* Anthony Arend, *"International Law and the Preemptive Use of Military Force;"* Marc Grossman, *"American Foreign Policy and the International Criminal Court;"* and Vesselin Popovski, *"International Criminal Court: A Necessary Step toward Global Justice."*

Those were useful readings, and quite timely as well. But some conceptual confusion found their way into the lesson plans, such as the assumption that the term *jus ad bellum* encompasses the whole of just-war theory. For 2004, the department responsible for teaching Course Two modified this lesson in light of our intended use of Anthony Hartle's book in Course One: e.g., the Course Two lesson will no longer need to provide an overview of just-war criteria. I hope we will instead focus here on issues regarding the U.S. *"preventive"* war in Iraq and the criteria needed to justify humanitarian military interventions, since there is no national consensus (let alone an international consensus) on either of those *jus ad bellum* topics.

Course Two also includes a lesson on National Values and National Interests, which examines the U.S. Declaration of Independence and Constitution, the 2002 National Security Strategy, and the Universal Declaration of Human Rights. That lesson will reinforce and enrich some ideas from Hartle's book that our students will have encountered in Course One.

The third core course, **Joint Processes and Landpower Development**, does not currently contain a lesson specifically dealing with ethics. Some ethical issues were hinted at this year in articles on mobilization and manning, such as the welfare of soldiers and families in connection with indefinitely long and dangerous assignments overseas, but they deserve to be explored in more depth. We should probably also examine ethical issues in defense acquisitions, which involve the negotiation of contracts with private corporations for weapons, supplies and services, and thus raise concerns about favoritism and conflicts of interest. Some of my colleagues have suggested that we would benefit more broadly from reading and discussing a book by Guy Adams and Danny Balfour, *Unmasking Administrative Evil.*

Our fourth core course, **Implementing National Military Strategy**, needs in my opinion to examine *jus in bello* rules of conduct in combat, the protection of civilians, the treatment of detainees from the *"war on terror,"* and the use of weapons like cluster bombs, depleted-uranium bullets, and earth-penetrating tactical nuclear weapons as *"bunker busters."* (Treatment of most of those issues was apparently lost when the just-war lesson was removed from Course One)

Elective Courses

In April 2004 I was teaching an elective course on **Ethics and Warfare**, assisted by two colleagues from other USAWC departments, Dr. Craig Nation and Marine Col. G. K. Cunningham. In previous years, this course relied heavily on Michael Walzer's *Just and Unjust Wars*, supplemented by articles on Vietnam and Kosovo that illustrated particular dilemmas in applying just-war criteria. I wanted to find a more comprehensive treatment of the origins and evolution of just-war thinking in Western political philosophy, and recently found an excellent one in Thomas Pangle and Peter Ahrensdorf, *Justice among Nations: On the Moral Basis of Power and Peace*. So our elective this year starts with a close reading of that book, along with some of my own essays on comparative religious ethics. Also, since most of this year's AWC students have not read Anthony Hartle's book, our elective will examine it next. Finally, the concerns expressed by Human Rights Watch (HRW) in their report, *Off Target: The Conduct of the War and Civilian Casualties in Iraq*, supplemented by comments from members of our student body who were witnesses to many of the events described by HRW will have to be examined. The attached course syllabus offers details on assigned and recommended readings. (Appendix 1)

Web Resources

I recently edited a website on **Just War Theory** in consultation with our Department of Distance Education and Horizons Unlimited, which will soon be available to both our resident and distance-learning faculty and students. It contains links to a wide range of articles, including classical sources of the theory in Cicero, Augustine, Aquinas, and Grotius, contemporary scholarly overviews of the field, case studies, and perspectives from non-Western cultures. I was able to draw in part from a list of recommended web resources that I had previously created for my undergraduate students.[70]

Future Directions

In my former life as a teacher of undergraduates, in addition to assigning materials on ethics and war from the perspectives of major world religions, pacifism vs. just war vs. total war etc., I also incorporated some powerful books in social psychology that may eventually find their way into courses at the USAWC. One was Michael Bilton and Kevin Sim, *Four Hours in My Lai*. That book provoked and enabled my undergrads to wrestle with questions about how the men of Charlie Company were capable of slaughtering hundreds of unarmed

[70] Online at http://home.earthlink.net/~davidlperry/weblinks.htm.

civilians at close range on March 16, 1968. Did they just turn out to be more callous or cruel than the average person? Was it rooted in the particular nature of the war in Vietnam? Was it due to bad training or leadership command of their Company? Are massacres like that inevitable in war? Are there tendencies in all or most people to do such things under similar conditions?

Bilton and Sim indicate that at least 14 written directives had been issued by the U.S. Army prohibiting the killing of noncombatants by its soldiers in Vietnam. So why were those directives not always followed? What factors tended to undermine them?

We first considered the tactics typically employed by the Vietcong: using villages as bases of operation, and fortifying them with mines and secret tunnels; dressing like civilians rather than in uniform; and attacking American patrols from inside villages. Such tactics tended to blur the line between combatant and noncombatant and threaten the lives of civilians.

But the authors also raise serious concerns about the training that Charlie Company received: Briefings on respect for noncombatants were relatively superficial and often not taken seriously. A company sergeant claimed that he trained his men to obey orders without question, and made it clear that disobedience would lead to serious consequences. He was also proud of how his men performed at My Lai. Lt. Calley testified that he assumed all orders to be legal, only to be questioned (if ever) after they were carried out. Racist comments about the Vietnamese and warnings that none of them could be trusted were frequently voiced informally by officers and returning vets, undermining concern for Vietnamese noncombatants.

Bilton and Sim further criticize some of the wider strategies and tactics used by the U.S. in that war. The strong emphasis from the top on "body count" meant that there were constant pressures on the ground to meet the objective of a high "kill ratio" of VC dead to American dead. And since there was little real concern by superior officers to verify whether dead Vietnamese were VC or not, American soldiers had an incentive to count *civilian* dead as VC, and thus not to be very scrupulous about putting their lives at risk. Moreover, the standard U.S. response to sniper or mortar fire from Vietnamese villages tended to be massive retaliation with artillery and bombing runs, leading to enormous collateral damage. We also frequently designated large areas as "free-fire zones," ostensibly warning local civilians to leave, but then subjecting those zones to indiscriminate strafing, shelling and bombing. These strategies and tactics almost inevitably led to violations of just-war criteria of noncombatant immunity and proportionality.

The authors also indicate that many experiences of Charlie Company after they arrived in Vietnam made some sort of massacre almost inevitable. A formidable

Vietcong battalion was known to operate in and around My Lai; their weapons and tunnels were discovered in the hamlets. All of Charlie Company's initial casualties were due to mines, snipers, and booby traps. But because they received no advance warnings about them from the villagers, they grew to distrust and hate them. Labeling people as *"VC sympathizers"* in effect led to their loss of noncombatant immunity.

Improper judgment and conduct were often exhibited by Charlie Company's officers. Capt. Medina ordered his men to kill some fishermen though they appeared to be unarmed. Prisoners were tortured and executed, some by direct orders of Medina and Calley. Other patterns of brutality by the soldiers started to develop. Civilians began to suffer beatings and random killings. Some of the men became serial rapists of local women, without being subject to any disciplinary action. And soldiers who were upset by these atrocities did not object to them openly or report them externally. Immediately prior to the massacre, Lt. Col. Barker apparently ordered My Lai to be *"neutralized"* without stipulating any limits on the treatment of noncombatants. And Medina was alleged by many of his men to have specifically ordered everyone in My Lai to be killed.

Since it is easy to be overwhelmed by the book's minute-by-minute account of the My Lai massacre, I have found it helpful to classify the range of behaviors exhibited by Charlie Company on that day in 1968: Some not only killed with enthusiasm, but also raped and mutilated their victims. Others killed efficiently and without emotion, as if shooting infants and old people were just like any other task they might be asked to accomplish. Some killed initially out of fear, but soon stopped in disgust or shame. Others went on killing even though they clearly felt anguished and sickened by it. Some showed compassion toward the wounded or warned villagers to run away. And some exhibited great courage in openly refusing to obey orders to slaughter clearly unarmed people, even when Lt. Calley threatened to shoot them for disobedience.

We then considered these questions: If we had been members of Charlie Company that day, what would we have been most likely to do? Would we have had the courage to disobey clearly illegal and immoral orders? Why did so few members of Charlie Company disobey?

Most of us want to believe that we could not bring ourselves to commit such atrocities. But the next book that was used in my undergraduate courses, Stanley Milgram's *Obedience to Authority*, suggests that we may be wrong to assume that. Milgram conducted a series of now-famous experiments in the 1960s in which subjects were led to believe that they were taking part in a study of the effects of punishment on learning; but they were actually being tested to see how many increasingly strong electric shocks they would inflict on a total stranger, simply because a skinny man in a white coat told them, *"The experiment requires that you continue."*

Now as you might imagine, Milgram's experimental subjects differed from Charlie Company in many ways: Most subjects were older. Some were women. The subjects were not encouraged to hate the person receiving the shocks, and had no prior reason to resent him. They received no special training to inculcate a habit of strict obedience. The man leading the experiment would seem to be much less authoritative than a military officer. And there were no penalties for refusing to obey particular commands or even quitting the whole experiment, unlike soldiers who might face court-martial or summary execution for disobedience in combat. But these differences make it all the more puzzling as to why most of Milgram's subjects obeyed completely, inflicting what they believed to be the highest possible shocks on a helpless *"learner."*

Milgram warns us not to infer from his experiments that most people deep down are really sadistic or cruel. In fact, one of his experiments showed that when people were allowed to choose the levels of shock to administer, they overwhelmingly tended to pick very light shocks. But Milgram's experiments are disturbing in showing that we tend to obey authority figures even when asked to do something that violates our conscience and induces extreme anxiety. This is something we ought to find profound and disturbing, because it suggests any one of us might have behaved like Paul Meadlo at My Lai, weeping openly as he mowed down dozens of unarmed people with his M-16. Milgram's findings also ought to be of great interest to military officers who seek to command soldiers in ways that minimize the likelihood of their committing atrocities in war.

Milgram helps us to understand some of the behaviors of Charlie Company, but not the most aggressive and cruel of them. That is why I followed that book with one by Richard Wrangham and Dale Peterson called *Demonic Males: Apes and the Origins of Human Violence*. Those authors draw some fascinating parallels between chimpanzee and human societies, suggesting that our pride, our desire to dominate others, and our tendencies to draw arbitrary divisions between *Us* and *Them* and to wage aggressive wars against Them, are deeply rooted in the evolution of our species. Although the authors don't believe that we are doomed to destroy one another, they effectively dispel the myth that human nature is originally and fundamentally peaceful. We cannot return to the Garden of Eden or a Golden Age of human civilization, because they never existed. We need to face the demon in our nature, and cultivate character traits, political institutions, and moral imagination to limit its destructiveness.

Military commanders need to be aware of the extremely aggressive violence that can emerge from otherwise gentle people under certain conditions, and avoid inducing such hatred of the enemy in their soldiers that they are unable to regard them as having basic rights. This is just as important at strategic, senior levels as it is in commanding a squad or platoon.

There is good reason to believe that our military commanders and training are much better today than they were 35 years ago in guarding against the risks of atrocities in war. During the Gulf War as well as recent wars in Afghanistan and Iraq, our soldiers have almost always shown admirable restraint and humanity toward enemy soldiers and civilians, even when enemy combatants shoot from behind civilians, hospitals, and mosques. One of my USAWC students this year wrote a fascinating paper on the degree to which respect for the laws of war is incorporated into Army training simulations at various levels (COL Kevin McClung, *"The Law of Land Warfare and Rules of Engagement: A Review of Army Doctrine and Training Methodologies."*) Although he determined that some improvements were warranted, he was pleasantly surprised by the quality of Army training in that regard. Another Army colonel who had served in Iraq during the 2003 invasion stated during a conference later that year at the Naval War College that his soldiers are trained so thoroughly in the laws of war, *"They know what the right thing to do feels like."* That suggests that Army training is wisely attempting to reinforce moral knowledge and critical reasoning with moral emotions and admirable character traits.

Finally, in June 2004, I had the opportunity to participate in a four-week symposium on just-war theory hosted by the U.S. Naval Academy and funded by the National Endowment for the Humanities. One of my goals for taking part in that symposium was to create a new elective for USAWC exploring the history of the just-war tradition, an understanding of which ought to enhance our students' ability to apply its principles appropriately in future military operations.

Conclusions

Ethics education at the U.S. Army War College is robust. It draws on the rich tradition of military ethics in the West generally and the United States in particular. Although as I have indicated there are a number of ways in which ethics pedagogy at AWC could be enhanced, those changes would be relatively small, not radical. The foundation is strong, but also amenable to improvements.

Appendix 1

U.S. ARMY WAR COLLEGE, CARLISLE BARRACKS, PA 17013-5050
ELECTIVE 160, ACADEMIC YEAR 2004-2005, TERM III
ETHICS AND WARFARE
FACULTY INSTRUCTORS: DR. DAVID PERRY,
ASSISTED BY COL G. K. CUNNINGHAM AND DR. CRAIG NATION

SECTION I: COURSE DESCRIPTION

1. OBJECTIVES

a. Develop an appreciation for the richness of moral thinking about war, especially in relation to the historical development of the just-war tradition.

b. Examine the core principles of the American professional military ethic (PME).

c. Assess the connections between the PME and the oath to defend the Constitution, the international laws of war, and the basic values of American society.

d. Enhance skills in applying ethical reasoning to issues and dilemmas facing contemporary strategic leaders.

e. In terms of Bloom's taxonomy of educational goals, this course primarily emphasizes Analysis, Synthesis, Evaluation, and Application, but it will also enhance Knowledge (of the history of relevant ethical theories) and Comprehension (of various ethical arguments related to warfare).

2. SCOPE

Military officers can encounter ethical dilemmas and challenges at any stage of their professional careers. But because the effects of one's decisions and actions tend to increase the further one advances in an organization, the ethical issues facing senior leaders can have momentous import. At any level within the military, mistakes in moral judgment can lead to unnecessary losses of human life; at higher levels, such mistakes can have truly catastrophic consequences. This course is intended to provide a basic foundation for sound ethical decision-making by strategic military leaders.

3. ELECTIVE REQUIREMENTS

a. Assigned materials are to be read carefully in advance of each class meeting. Class sessions will primarily focus on discussion of those readings, occasionally supplemented by faculty lectures.

b. Students will complete a 5- to 7-page double-spaced paper on an aspect of just war or military ethics relevant to senior military leadership. Normally, students will choose a topic prompted from the assigned reading for further exploration and application. *Suspense for the paper will be 1600 on 28 May 2004.* See also Appendix A for additional guidelines for your paper.

c. Avoid missing class sessions; ideally seek Faculty Instructor approval for absences ahead of time.

SECTION II: CLASS DESCRIPTIONS

1. **Introduction**. What is ethics? What's the value of studying ethics? Is morality objective? Are any moral principles, rules, or values absolute? What are some objective ethical principles? What are some guidelines for ethical deliberation and action? How should we categorize ethical issues regarding war and the military? Even if we could agree on a set of ethical principles that would apply objectively to soldiers, are there psychological and social challenges in upholding high standards of conduct in warfare?

Required Readings: None. (Faculty will lecture and facilitate discussion.)

2. **War and Political Philosophy I:** In this session we'll begin our exploration of the roots of western ethical thinking about international relations in general and war in particular, with special attention to the thoughts of ancient Greek writers Thucydides, Plato and Aristotle.

Required Reading: Thomas Pangle and Peter Ahrensdorf, *Justice among Nations: On the Moral Basis of Power and Peace*, 1-50.

3. **War and Political Philosophy II:** We will continue our historical survey via an examination of the secular ideas of the Stoics and the Roman jurist Cicero. We will also see how early Christian writers initially upheld a pacifist ethic, while their successors contributed to what would come to be known as the just-war tradition.

Required Reading: Pangle and Ahrensdorf, 51-124.

4. **Modern Realism:** During the Renaissance and Reformation period in Europe, some political theorists such as Machiavelli and Hobbes adopted a skeptical view of ethics and a critical attitude toward the perceived naïveté of their philosophical predecessors. We will explore their startling ideas, as well as the thoughtful retorts of Locke and Montesquieu.

Required Reading: Pangle and Ahrensdorf, 125-161.

5. **Modern Idealism:** In this session we'll examine the work of Hugo Grotius, one of the most important figures in the just-war tradition and a founder of modern international law. We will also trace his influence on Pufendorf, Wolff, Vattel, Burke, Rousseau and Kant.

Required Reading: Pangle and Ahrensdorf, 162-217.

6. **Contemporary Realism and Neorealism:** With the help of 20th-century political theorists Morgenthau and Waltz, we will discuss what we may sensibly take from the rich traditions of western realism and idealism in order to evaluate armed conflicts today.

Required Reading: Pangle and Ahrensdorf, 218-263.

7. **American Military Ethics and the Laws of War**: What are the core values of the *American* military ethic? What does it mean to "support and defend the Constitution of the United States against all enemies, foreign and domestic"? What are some specific ethical implications of "Duty, Honor, Country"? How do the international laws of war relate to the American military code of ethics?

Required Reading: Anthony Hartle, *Moral Issues in Military Decision Making*, 36-84.

8. **Military Ethics re: the Values of American Society**: How does the American military ethic compare with the values of wider society, such as freedom, equality, individualism and democracy? Is the military ethic consistent with those values, or in tension with them?

Required Reading: Hartle, 85-128.

9. **Applying Jus in Bello Principles to Contemporary Military Operations I:** How did the behavior of Iraqi forces during major combat operations in 2003 complicate the efforts of American and allied military forces to uphold high ethical and legal standards of conduct? How well did we do in our air campaign to minimize civilian casualties?

Required Reading: Human Rights Watch, *Off Target: The Conduct of the War and Civilian Casualties in Iraq*, 1-63.

10. **Applying *Jus in Bello* Principles to Contemporary Military Operations II:** How should we evaluate ethically the conduct of our ground campaign in Iraq? Are there lessons that we can learn from targeting or weaponeering decisions that produced excessive civilian deaths and injuries, even if unintentionally?

Required Reading: Human Rights Watch, *Off Target*, 64-126.

Guidelines for Your Paper

Topics and Sources. The paper must deal in some way with ethics and warfare, and incorporate ideas from at least two sources outside of the assigned course readings. (You may use our assigned readings, too, but you aren't required to do so.) Newspaper or magazine articles may be cited, but your main sources should be scholarly. Since the paper is to be fairly short (5-7), it's important to focus your topic narrowly in order to do it justice. You should elicit feedback from FIs on your paper topic soon, so that they can steer you away from inappropriate subjects and toward useful research sources.

You may use the paper to argue in support of one side of an ethical issue. If so, then you should not only find substantial arguments in support of that position, but also tackle substantial arguments against that position. In other words, try to show that your position successfully withstands strong criticism. Alternatively, you might want to explore an ethical issue without taking a firm stance on either side. In that case also, find substantial arguments that produce conflicting conclusions, and analyze their strengths and weaknesses. Do not choose a topic where the only credible ethical arguments favor one side: e.g., do not write a paper on why intentionally slaughtering clearly unarmed civilians would be morally questionable.

Format. Your paper should be typed and double-spaced, have one-inch margins on all sides, and employ a 12-point font in a style like Times Roman or Ariel (but not Courier). Do not add covers or a title page; simply provide your name and the title at the top of the first page of the body of your essay, and staple the pages together in the upper left corner.

Editing. Your FIs are willing to comment on a draft of your paper at least two weeks prior to the due date for the final product. Check carefully to eliminate all errors in spelling and grammar, even in your draft. FI energies should be spent analyzing the quality of your arguments rather than correcting mistakes that you could easily have caught yourself.

Writing with Integrity. Hugo Bedau wrote in *Thinking and Writing about Philosophy*, p. 141: "Writers plagiarize when they use another's words or ideas without suitable acknowledgement. Plagiarism amounts to *theft*-theft of language and thought. Plagiarism also involves *deception*... [Plagiarism] wrongs the person from whom the words or thoughts were taken and to whom no credit was given; and it wrongs the reader by fraudulently misrepresenting the words or thoughts as though they are the writer's own." In addition, plagiarism is also unfair to your fellow students, since in contrast to their honest and hard work, it involves almost no effort to steal another author's words.

Finally, when you plagiarize you also cheat yourself: first, by not developing the discipline and diligence to research, write and edit well; second, because taking credit for other people's ideas will induce outrage and resentment against you; and third, because a habit of plagiarism can end your career and destroy your reputation. Witness what happened to the prominent historians Doris Kearns Goodwin and the late Stephen Ambrose when substantial parts of their books were revealed to have been plagiarized: their reputations were ruined and they stopped appearing on TV as credible experts.

To avoid plagiarism, you must cite your sources everywhere in your paper where you use their ideas, and not only when you quote them directly, but also where you paraphrase their points in your own words. In general, you should only use direct quotes when you find the authors' wording to be especially effective. But your paraphrasing or summaries of authors' points should be thorough: it is not fair to an author to change only a couple of words in a paragraph of his or hers, and then imply (by not using direct quotes) that the paragraph is entirely your own prose. (It might help to imagine the author reading over your shoulder.)

Some additional practices are unethical for similar reasons:

1. You may not submit a paper that you did not write yourself, whether given to you by a sympathetic friend, purchased on the web, or written by a paid "tutor." Although (unlike plagiarism) this would not represent a case of theft, it would nonetheless be unfair to your peers, and would also perpetrate a fraud on your FIs.

2. You may not submit a paper that is substantially similar to one that you wrote for another class, since both the FIs and your fellow students rightly assume that you will be doing new research for this course.

List full references alphabetically at the end of your paper, using a consistent scholarly format. If you use endnote references and include all of your source data in them, there's no need to add a separate bibliography as well.

References

Aquinas, Thomas: Is It Always Sinful to Wage War? *Summa Theologica* II-2,

Augustine: Letter to Marcellinus. 412, http://www.newadvent.org/fathers/1102138.htm.

Bainton, Roland (1960): *Christian Attitudes Toward War and Peace*. Nashville, Abingdon.

Bartholomeusz, Tessa (1999): In Defense of Dharma: Just-War Ideology in Buddhist Sri Lanka. *Journal of Buddhist Ethics*. http://jbe.gold.ac.uk/6/bartho991.htm.

Bilton, Michael; Sim, Kevin (1992): *Four Hours in My Lai*. New York, Penguin Books.

Brinsfield, John (1998): Army Values and Ethics: A Search for Consistency and Relevance. *Parameters*, Autumn,

Brown, Chris (2003): Self-Defense in an Imperfect World. *Ethics and International Affairs*.

Byers, Michael (2003): Letting the Exception Prove the Rule. *Ethics and International Affairs*.

Carlino, Michael (2002): The Moral Limits of Strategic Attack. *Parameters*, Spring.

Coady, C.A.J. (2002): *The Ethics of Armed Humanitarian Intervention*. Washington DC, U.S. Institute of Peace.

Coates, Anthony (1997): *The Ethics of War*. New York, Manchester University Press.

Cohen, Marshall, et al. (eds.) (1974): *War and Moral Responsibility*. Princeton: Princeton University Press.

Cook, Martin (2000): Immaculate War: Constraints on Humanitarian Intervention. *Ethics and International Affairs*.

Coppieters, Bruno; Fotion, Nick (eds.) (2002): *Moral Constraints on War*. Lanham, MD, Lexington Books.

Dower, John (1986): *War without Mercy: Race and Power in the Pacific War*. New York, Pantheon Books.

Dworkin, Ronald (2003): Terror and the Attack on Civil Liberties. *New York Review of Books*, 6 November.

Eikenberry, Karl (1996): Take No Casualties. *Parameters*, Summer,

Erasmus, Desiderius (1989): The Complaint of Peace. *The Praise of Folly and Other Writings*. New York, W.W. Norton.

Firestone, Reuven (1999): *Jihad: The Origin of Holy War in Islam.* New York, Oxford University Press.

Ford, John (1944): The Morality of Obliteration Bombing. *Theological Studies*, September.

French, Shannon (2003): *The Code of the Warrior: Exploring Warrior Values Past and Present*. Lanham, MD, Rowman and Littlefield.

Gong, Ted (1999): China: Tradition, Nationalism, and Just War. Strategy Research Project. Carlisle, PA, U.S. Army War College.

Grotius, Hugo (1625): *The Rights of War and Peace*.

Gutman, Roy; Rieff, David (1999): *Crimes of War*. New York: W.W. Norton.

Hartigan, Richard (1983): *Lieber's Code and the Law of War*. Chicago, Precedent.

Hartle, Anthony (2004): *Moral Issues in Military Decision Making.* Second edition. Lawrence, KS, University Press of Kansas.

Harvey, Peter (2000): War and Peace. *An Introduction to Buddhist Ethics*. New York, Cambridge Univ. Press.

Hinman, Lawrence: Ethics of War, Peace, and Terrorism. Miscellaneous documents and videos, http://ethics.acusd.edu/Applied/Military/index.asp.

Hoffman, Bruce (2002): A Nasty Business. *Atlantic Monthly*, January.

Holmes, Arthur F. (ed.) (1975): *War and Christian Ethics*. Grand Rapids, Baker Book House.

Howard, Michael, et al. (eds.) (1994): *The Laws of War: Constraints on Warfare in the Western World*. New Haven, Yale University Press.

Human Rights Watch (2001): *Cluster Bombs in Afghanistan*. New York.

Ignatieff, Michael (2004): A Year of Living Dangerously. *New York Times*, 14 March.

International Committee of the Red Cross: International Humanitarian Law. Miscellaneous documents dated 1856 to the present, http://www.icrc.org/ihl.nsf/WebFULL?OpenView.

Johnson, James Turner (1984): *Can Modern War Be Just?* New Haven, Yale University Press.

Kagan, Robert; Kristol, William (2004): The Right War for the Right Reasons. *Weekly Standard*, 23 February.

Kant, Immanuel (1795): *Perpetual Peace: A Philosophical Sketch*. http://www.constitution.org/kant/perpeace.htm.

Keen, Maurice (1965): *The Laws of War in the Late Middle Ages*. London: Routledge and Kegan Paul.

Kelsay, John (1993): *Islam and War: A Study in Comparative Ethics*. Louisville, Westminster; John Knox Press.

Klostermaier, Klaus (1996): Himsa and Ahimsa Traditions in Hinduism. *The Pacifist Impulse in Historical Perspective*. Toronto, University of Toronto Press.

Knapp, Michael (2003): The Concept and Practice of *Jihad* in Islam. *Parameters*, Spring, http://carlisle-www.army.mil/usawc/Parameters/03spring/knapp.htm.

Krauss, Eric; Lacey, Mike (2002): Utilitarian vs. Humanitarian: The Battle over the Law of War. *Parameters*, Summer, http://carlisle-www.army.mil/usawc/Parameters/02summer/lacey.htm.

Lichtenberg, Judith (2001): The Ethics of Retaliation. *Philosophy and Public Policy Quarterly*, Fall. http://www.puaf.umd.edu/IPPP/reports/fall,%20vol%2021,%202001/222984.pdf

Locke, John (1689): Of Conquest. *Two Treatises of Government*, http://www.cooperativeindividualism.org/locke_on_conquest.html.

Luban, David (2002): The War on Terrorism and the End of Human Rights. *Philosophy and Public Policy Quarterly*, Summer, http://www.puaf.umd.edu/IPPP/reports/vol22sum02/Summer%202002%20issue.pdf.

Luther, Martin (1526): Whether Soldiers, Too, Can Be Saved. http://www.lccabc.ca/Resources/War/WhetherSoldiers.pdf.

Maass, Peter (2003): 'Good Kills.' *New York Times*, 20 April.

Milgram, Stanley (1974): *Obedience to Authority*. New York, Harper and Row.

Miller, Richard (ed.) (1992): *War in the Twentieth Century: Sources in Theological Ethics*. Louisville, Westminster; John Knox Press.

Miller, William (2000): *The Mystery of Courage*. Cambridge, MA, Harvard University Press.

Moore, Jonathan (ed.) (1998): *Hard Choices: Moral Dilemmas in Humanitarian Intervention*. Lanham, MD, Rowman and Littlefield.

Nardin, Terry (ed.) (1996): *The Ethics of War and Peace: Religious and Secular Perspectives*. Princeton, Princeton University Press.

Nichols, Thomas (2003): Just War, Not Prevention. *Ethics and International Affairs*.

Norman, Richard (1995): *Ethics, Killing, and War*. New York, Cambridge University Press.

Orend, Brian (2002): War. *Stanford Encyclopedia of Philosophy*, May, http://plato.stanford.edu/entries/war/.

Pangle, Thomas Ahrensdorf, Peter (1999): *Justice Among Nations: On the Moral Basis of Power and Peace*. Lawrence, KS, University Press of Kansas.

Perry, David (2003): Using Shakespeare's *Henry V* to Teach Just-War Principles, http://home.earthlink.net/~davidlperry/henryv.htm.

Peters, Rudolph (1996): *Jihad in Classical and Modern Islam*. Princeton, Markus Wiener.

Potter, Ralph B., Jr. (1969): *War and Moral Discourse*. Richmond, John Knox Press.

Power, Samantha (2001): Bystanders to Genocide: Why the United States Let the Rwandan Tragedy Happen. *Atlantic Monthly*, September.

Ramsey, Paul (1968): *The Just War: Force and Political Responsibility*. New York, Scribner.

Regan, Richard (1996): *Just War: Principles and Cases*. Washington, DC, Catholic University Press.

Rohr, John (1971): *Prophets Without Honor: Public Policy and the Selective Conscientious Objector*. Nashville, Abingdon.

Sapolsky, Harvey; Shapiro, Jeremy (1996): Casualties, Technology, and America's Future Wars. *Parameters*, Summer, http://carlisle-www.army.mil/usawc/Parameters/96summer/sapolsky.htm.

Schlissel, Lillian (ed.) (1968): *Conscience in America: A Documentary History of Conscientious Objection in America, 1757-1967*. New York, Dutton.

Steger, Manfred (2000): *Gandhi's Dilemma: Nonviolent Principles and Nationalist Power*. New York, St. Martin's Press.

Swift, Louis (1983): *The Early Fathers on War and Military Service*. Wilmington, Michael Glazier.

Tuck, Richard (1999): *The Rights of War and Peace: Political Thought and the International Order from Grotius to Kant*. New York, Oxford University Press.

Verkamp, Bernard (1993): *The Moral Treatment of Returning Warriors in Early Medieval and Modern Times*. Scranton, University of Scranton Press.

Victoria, Brian (1997): *Zen at War*. New York, Weatherhill.

Vitoria, Francisco (1991): On the Law of War. *Vitoria: Political Writings*. New York, Cambridge University Press.

Wakin, Malham (ed.) (1986): *War, Morality, and the Military Profession*. Second edition. Boulder, CO, Westview Press.

Walters, LeRoy (1974): Historical Applications of the Just War Theory. *Love and Society: Essays in the Ethics of Paul Ramsey*. Atlanta, Scholars Press.

Walzer, Michael (2000): *Just and Unjust Wars: A Moral Argument with Historical Illustrations*. Third edition. New York, Basic Books.

Wrangham, Richard; Peterson, Dale (1996): *Demonic Males: Apes and the Origins of Human Violence*. New York, Houghton Mifflin.

Yoder, John Howard (1996): *When War Is Unjust: Being Honest in Just-War Thinking*. Second edition. Maryknoll, NY, Orbis.

Andreas KASTBERGER, National Defense Academy, Vienna

EXPERIENCES WITH ETHICS EDUCATION IN THE NON-COMMISSIONED OFFICER'S ACADEMY OF THE AUSTRIAN ARMED FORCES

Introduction

As for ethics, people normally think that such a topic only interests academics. Concerning the ethical problems of armed forces, civilians are sometimes astonished by the fact that our army trains all its soldiers (not only the commanding officers) in dealing with ethical and moral problems. Education in ethics for NCOs is an initiative which has to be regarded as a challenging approach. It should be clear that our soldiers must know how to respond to moral questions. Therefore squad leaders and platoon commanders' knowledge of ethics has to be brought to a minimum level in order to qualify these leaders for their jobs in the field.

One of the questions that always arises when teaching ethics at the NCO's academy is: *"What is the practical use of ethics education, e.g. for a squad leader doing his job at a check point in a peace support operation (PSO)?"* From their point of view, theory and practice in general seem to be contradictory. So it is the responsibility of the teacher to show them that theory should be regarded as the necessary basis for practice. Reaching that goal is much easier with students in the rank of Warrant Officers, especially when these soldiers are able to put forward practical problems out of their mission experiences. Honestly speaking, it is quite a challenge to explain the practical use of moral theory to future NCOs who tend to be about 20 years old. Taking the high responsibility of commanders into consideration, beginning at the lowest level of a squad or section, the necessity of ethics education even for these leaders should be without question.

Therefore the main idea of training ethics at the NCO's academy is to combine theoretical basic information and practical questions out of the soldiers' experiences. The input of theory, in general, should not be too complex, but that does not mean it has to be oversimplified. Taking care of the soldiers' needs in daily work or in specific missions may be the key to success in the end. Otherwise the teacher would fail because course participants could not understand the content of the theoretical briefings and the importance of ethics for the profession of a soldier.

The training system for NCOs of the Austrian Armed Forces

To give a brief overview about the general course system and methods implemented for the NCO's training, it is first of all worth pointing out that the courses and seminars are conducted in three levels. Level 1 is the basic training for future NCOs. The students are to be prepared in that course to take over the leadership and the responsibility for the education of a squad (a team consisting of 8 - 12 soldiers). Level 2 is the advanced training for platoon commanders, who have to lead and train a platoon (4-5 squads). Level 3 – the staff for the NCO's further training has to cover the widest field of educational tasks, including a course on staff duties at brigade level, leadership techniques, train-the-trainer seminars, time management workshops, foreign language training (English) or seminars on civics and ethics. The NCO's academy is only responsible for the general training and education of the non-commissioned officers in the AAF. The job-related courses are conducted by eleven schools of special branches, e.g. infantry, tank, artillery, or engineers.

Education on military ethics for NCOs in Austria

As already mentioned above, education in ethics is part of the curricula in the NCO's training in each of the three levels. In the NCO's basic and advanced trainings the contents of the lessons stick to general rules of a soldiers' engagement with a direct link to practical problems. Inputs on moral questions are taught by an expert on ethics in one day, conducted as an intensive seminar. The time period to be reserved for ethics questions is extended in the highest level of the NCO's training. The courses, seminars, and workshops for high ranking NCOs can be attended on a voluntary basis. The ethics seminar can be found in the training program once a year (usually in autumn / duration: 3 days). The main idea of these seminars is to show some cases taken out of the seminar schedules of the last two years.

Seminars on ethics for NCOs in the Austrian Armed Forces

The major challenge of these seminars is – as previously pointed out – the combination of a theoretical background on military ethics as a part of practical skills to be used in different kinds of military operations and in civil-military relations, as well. Each seminar has its own topic, mostly worked out by the participants giving feedback and handing over proposals for future programs. The first seminar was held in 1999 with a lot of organizational problems. The idea of the interdisciplinary approach came up in 2000, and it became the main characteristic of the yearly seminar in the NCO's academy. The students should come to understand moral problems by being confronted with the results of research in different academic branches.

The seminar in 2001 should be taken as the first case in point. The question put forward to the students was "should the powers of discernment in moral aspects be regarded as a basic skill for coping with military missions in general, or not? It turned out to be necessary to start with an overview about ethics, morals, and customs in modern society, especially to define the terms to be used in further discussions. That part was taken over by a theologian. Then the focus was put on a developmental psychological approach based on the model of ethical development by Lawrence Kohlberg. The main interests of the seminar participants could be found in problems which usually arise in peace support operations. According to the students the seminar was a great success, but the major criticism in a negative sense raised the above mentioned necessity to increase the possibilities of the practical use of the theoretical inputs.

The internal evaluation of the seminar resulted in an even more practically useable schedule. In 2002, the academy promoted a seminar on civil-military co-operation (CIMIC), the problem of the return and integration of refugees and minorities in crisis areas as one of the tasks of the CIMIC staffs in PSO. The moral aspect of that topic was obvious: evaluating the different approaches of international and national organizations dealing with the question of refugees' returning in missions abroad. Can the return of refugees be seen as a human right? In practice, you can find different opinions, at least different focuses, between organizations like UNHCR (United Nations High Commissioner for Refugees) or the military, which is normally responsible for the safety and security of all the people in the mission area. The example of Kosovo with the problem of the return and integration of Kosovo Serbs was taken and analyzed by several experts. For the CIMIC staff in the KFOR-mission (Kosovo Forces) it is of high importance to get some background on religious basics. What should a Christian do in a dilemma situation as a KFOR-soldier, being confronted with other organizations that claim KFOR does not stick to human rights by trying to protect people and advising others to be more careful with the acceleration of return programs for Serbs, who are still seen as former war criminals by the Albanian majority? What are the legal basics of the international law on the right of people to return to their places of origin? How can the experiences of KFOR-soldiers in the CIMIC branch be used for further steps in civil-military relations? And finally, what is the responsibility of the members of the NCO corps in such a mission? How can the importance of NCOs in missions abroad, especially in CIMIC tasks, be increased in the future, and what are the prerequisites to do that? The academy succeeded in convincing some professional trainers to join the seminar – a theologian and expert on military ethics, a staff scientist who specialized on international law, and two staff members of the NCO's academy. One of these two teachers was a Warrant Officer out of the section for civic education; the other one was the seminar director, additionally with mission experiences as a CIMIC staff officer in Kosovo, 2001.

Experiences and future aspects

All in all, the seminars on military ethics of the last years can be seen as a great success. The importance of theoretical basics with a main focus on the practical use of ethics for the soldier's profession should be kept as a guideline for future planning in case of ethics training for non-commissioned officers. The motivation of NCOs, especially in the rank of Warrant Officers with more than ten years of service in the armed forces, to deal with moral problems and to gain knowledge and skills in order to solve them, is generally very high. This is a chance the army has to take because the responsibility of non-commissioned officers in modern missions abroad has steadily increased.

A basic education in ethics is not a guarantee for the success of soldiers carrying out their tasks in the battlefield or in a PSO scenario, but the deployment of soldiers without a minimum level of moral attitude in a mission would, at least in the long run, be the guarantee for failure.

Moral Challenges
arising from Asymmetric Warfare

Hermann JUNG, Austrian Armed Forces (ret.), Vienna

PREPARING FOR ASYMMETRY – JOINT VISIONS 2010 AND 2020 – AN ORGANIZATIONAL LEARNING PERSPECTIVE

Historically, power nations have considered victory in war as paramount. What was true for Europe in the 19th century appears now true for the US. They did not set out just to fight the nation's wars, they planned to fight and win the nation's wars. The whole point of Joint Visions (JV) 2010/2020 is to provide an adequate roadmap for US military forces to prepare to win – decisively. Power nations inherently tend to be more offensive in nature. And in conflicts that fall below Joint Visions' warfighting threshold, power nations can control situations, can seize initiative, and achieve specific objectives; but the last 10 years of protracted operations have forced them to redefine victory on several levels.

Reasons for global players to adopt new patterns

(I) Overwhelming military capability alone does not effectively deter regional or internal conflicts from igniting or becoming entrenched as protracted cycles of violence. Threats of military intervention may play a role in major power state decision-making, but it has not proven to be an effective conflict prevention mechanism among lesser powers and toward substate warring factions. There is little concrete evidence that the existence of a powerful military can deter nonstate actors from pursuing their objectives. Indeed, overwhelming force, globally deployed, offers lucrative targets of opportunity to those dedicated to demonstrating weaknesses and vulnerabilities of global powers. Those actors circumvent military operations through asymmetric approaches.

(II) There is no certainty even for overwhelming forces to achieve full victory or win against the kind of conflicts that characterize today's global security environment. A special problem is posed by civil wars, insurgencies, or cross-border incursions where no *"enemy"* force is clearly identified. Deployed forces operating under present conditions can control events, situations, and terrain by exerting constant pressure and perpetual presence. However, the threshold in these types of conflict inherently falls below requirements for *"the deployment of overwhelming forces,"* so *"victory"* can not be guaranteed.

(III) With no direct strategic concepts dedicated to overcoming the potential for asymmetric approaches, global players and their overwhelming forces are left to defend, adapt, or develop operational concepts in an ad hoc manner. Because of their close association to *"warfare on the cheap,"* or *"warfare without rules,"* or just plain free-for-alls, asymmetric dynamics also tend to rate a lower priority in terms of devising effective response mechanisms other than defensively.

(IV) Based on current or projected defense spending, allies of the US have not made conscious decisions to keep pace with the advances in lethality, precision, Command, Control, Communications, Computers, Intelligence, Surveillance and Reconnaissance (C4ISR), etc. There are few indications that allies are in a position or are willing to expend the resources necessary to become full partners under the doctrine of overwhelming force as it is defined in JV 2010/2020. This poses unique challenges in a strategic environment where stated policy is one that dictates to act unilaterally if necessary, but will always prefer to operate in concert with allied partners. Allies may also look more to building their own capabilities with their own vision and purpose, which may in the end create more problems than they solve in terms of interoperability. They may also lead the way in terms of developing operational concepts to manage lower threshold conflicts for which U.S. forces and operational concepts may not be able to *"pull back on the throttle"* enough to adjust effectively.

(V) Finally, the dominant force posture may very likely be driving military developments by potential adversaries in response to their perception that they may be targets of the *"Shape, Prepare, and Respond"* strategic conception. In other words, the forces and capabilities envisioned in JV 2010/2020 are also making asymmetric approaches more and more attractive. This *"perception-reaction"* pattern to US military visions is not an argument to *"dumb down"* U.S. forces in an effort to *"level the playing battlefield;"* rather, it is an effort to show the need for a more inclusive strategy that incorporates asymmetric dynamics as equally significant factors for consideration by force planners, strategists, and political decision-makers.

Today's Transatlantic Problem

Today's transatlantic problem, in short, is a power problem. American military strength has produced a propensity to use that strength. Europe's military weakness has produced a perfectly understandable aversion to the exercise of military power. Indeed, it has produced a powerful European interest in inhabiting a world where strength does not matter, where international law and international institutions predominate, where unilateral action by powerful nations is outlawed, where all nations regardless of their strength have equal rights and are equally protected by commonly agreed-upon international rules of behavior. (Kagan, 2003)

This is why in recent years a principal objective of European foreign policy has become, as one European observer puts it, the "multilateralizing" of the United States. (Everts, 2001) It is not that Europeans are teaming up against the American hegemon, as realist theorists would have it, by creating a countervailing power. After all, Europeans are not increasing their power. Their tactics, like their goals, are the tactics of the weak. They hope to constrain American power without wielding power themselves. In what may be the ultimate feat of subtlety and indirection, they want to control the behemoth by appealing to its conscience.

Even when Americans and Europeans agree on the kind of world order they would strive to build, they increasingly disagree about what constitutes a threat to that international endeavor. Indeed, Europeans and Americans differ decisively in their evaluation of what constitutes a tolerable versus an intolerable threat. This, too, is consistent with the disparity of power.

A far better explanation of Europe's greater tolerance for threats is, once again, Europe's relative weakness. Europe, precisely because it is weak, actually faces fewer threats than the far more powerful United States.

What does it mean to rely on allies who have a different threat perception and who are placed in the asymmetric corner by US power projection?

At first, it will be important to show, through the lens of JV 2010/2020, what the U.S. forces expect from their allies in future conflicts, since history, strategy, and recent experience suggest that the US usually works in concert with her friends and allies in almost all operations.

The primary task of armed forces will remain to deter conflict - but, should deterrence fail, to fight and win the nation's and alliance's wars. In addition, armed forces should expect to participate in a broad range of deterrent, conflict prevention, and peacetime activities. Joint and combined coalition training and exercises have improved transatlantic interoperability and understanding of the strengths of each individual service as well as allies and coalition partners. From the individual warfighter to large multinational forces, this systematic approach has enabled forces and soldiers to hone their skills in practice many times before ever having to perform actual combat missions. These training innovations must be sustained.

As these visions are implemented, to acknowledge strong leadership, warfighting skills, and innovative thinking will be central to developing the detailed requirements. Our organizational climate must reward *critical thinking, foster the competition of ideas, and reduce structural or cultural barriers to innovation.* Both in peace and war, the creative talents of our men and women provide us a critical advantage over those who would consider challenging the U.S. or its allies.

The Need for an Integrated Vision 2010 and 2020

Asymmetry cannot be treated as a *"lesser included case."* US forces can, and must, embrace asymmetry as a challenge for which there are effective solutions.

These solutions may not look quite like the traditional way of waging wars as we have known it, but effective ways and means of including asymmetry as an integral aspect of the joint vision can be developed. Operational concepts must accommodate the management of change to allow transition, or evolution, from existing ways of doing business to institutionalize the multidimensional challenges of asymmetry. The objective result will be a transformed military *"means"* to address new *"ways"* of conflict to achieve *"ends"* in the 21st century.

The intent of this *transition* is to develop the appropriate processes, refine doctrine, organization, training, materials, leadership, personnel, and facilities; (DOTMLPF), and institutionalize consideration of asymmetric dynamics as a relevant, legitimate concern at all levels among all disciplines.[71] The end-state of transformation would essentially fully integrate the asymmetric domain into the JV2010 and JV2020 construct. The transformation would result in the problem of asymmetry becoming manageable and, with some exceptions, a normal part of military operations today and tomorrow. Transformation would institutionalize understanding of the scope, significance, and impact of the set of problems, and provide the necessary tools and processes to support planning and operational responses to asymmetry. Some asymmetric threats will still fall outside the military's ability to anticipate or will require a unique response outside the normal parameters of strategic planning and operational processes.

Preparing forces to operate in a world where asymmetry appears to be the only logical option for adversaries requires some significant and innovative adaptations to training and education methods. It essentially requires forces trained to face a multiplicity of asymmetric strategies from a host of players in a vast array of complex environments, all intended to foil traditional regular forces' ability to deploy and fight as designed. It can no longer rely on training to face a like-minded enemy as a successful approach for military preparation against asymmetry because the world of conflict is no longer a simple blue versus red equation, but increasingly a blue versus red versus green versus orange mélange.

The dynamic nature of joint operations in the 21st century battlespace will require a continued emphasis on developing strong *leadership* skills. (Joint Vision, 2010: 28) While everything possible to leverage the power of advanced technologies needs to be pursued, there are inherent limitations. Confronting the inevitable friction and fog of war against a resourceful and strong-minded adversary, the human dimension including innovative strategic and operational thinking and strong leadership will be essential to achieve decisive results. Effective leadership provides the greatest hedge against uncertainty. US Forces

---

[71] Strategic plan for Transforming DoD Training, March 2002

and their allies will build upon the enduring foundation of functional expertise, core values, and high ethical standards.[72] Future leaders at all levels of command must understand the interrelationships among military power, diplomacy, and economic pressure, as well as the role of various government agencies and non-governmental actors, in achieving security objectives. They will have to require a sophisticated understanding of historical context and communication skills to succeed in the future. The evolution of command structures increased pace and scope of operations and the continuing refinement of force structure and organizations will require leaders with knowledge of the capabilities of all four services. Without sacrificing their basic service competencies, these future leaders must be schooled in joint operations from the beginning of their careers. This leadership development must begin with a rigorous selection processes and extend beyond formal education and training. Experience in a variety of progressive assignments must stress innovation, dealing with ambiguity, and a sophisticated understanding of the military art. In short, all leaders must demonstrate the very highest levels of skill and versatility in ever more complex joint and multinational operations.

Multidimensional Thinking and Acting[73]

Some experts maintain that the current U.S. approach to military operations – strategic, operational, and tactical – as it is true for all Western Democracies is too linear for today's contemporary operating environment. Future warfighters must move beyond linear thought and action to a realm of thinking and acting that recognizes and accepts paired yet opposite ideas and actions: *"Look before you leap"* and at the same time understand that *"he who hesitates is lost."*

Traditionally, a nation-state attacks another with military force, and the response is rather predictable. Today, the qualities of nation-states are no longer required to initiate attacks; attacks may not even have traditional military qualities; and prediction is just not as calculable as before.

"Asymmetric warfare" shapes the future. Some postulate that the problem of asymmetric conflict is at the strategic level. The underlying assumption of western military thinking and acting is that we can address asymmetric problems with hierarchically directed linear thinking. The argument is that the strategic, operational, and tactical paradigm exists because it enables the military to adapt through echelonment.

---

[72] For broader information see FM 22-100, Army Leadership.

[73] For broader information see Military Review January-February 2002, Col. Christopher R. Paparone, U.S. Army, and James A. Crupi. These authors call this method of thinking and acting *"Janusian"* that would mean thinking and deciding in (only) two directions. But they propose at the end at least a more directional method, therefore it is better to speak of *"multidimensional"*, the author.

The danger is that structure ends up driving response, instead of needed capabilities and values driving organizational response.

A **INTUITION** **FEELING**	**B** **INTUITION** **THINKING**
SENSING **JUDGING** **D**	**SENSING** **PERCEIVING** **C**

Fig. 1: Four-way interactive thinking

Linear thinking and acting is inadequate, given the nature of conflict in the postmodern era. The preferred new paradigm must have several characteristics that set it apart from the traditional strategic paradigm.

It must have a *fractal quality*[74] that allows us to take simultaneous full spectrum looks at human information processing. What we really need is an alternative paradigm that gives us a new and better way of thinking and acting. The new approach should provide a range of insights that enable commanders to instantly conceptualize a pattern of multidimensional possibilities that lead to breakthrough concepts and values.

It should not reject traditional levels of analysis such as strategy, operations, and tactics but should put them as secondary concepts. Concepts such as simultaneity of paradoxes, comprehending activities in multiple time orientations, and embracing environmental complexity (unpredictability) as a normal condition must be emphasized.

Some may argue that increasing the speed of linear decision-making will address the chaotic nature of unfolding events, but that is not the case. Above all else, linear thinking and acting disallow the existence of contradiction. The proposed alternative suggests that information processing is paradoxical, considers multiple time orientations, is nonlinear, and is more than two-dimensional.

This new way of thinking goes beyond rational thinking. Multidimensional thinking makes it possible to make sense of the postmodern world in an almost circular, interconnected, interdependent way and, as a result, represents a more

[74] Extremely irregular curve or shape where any suitable chosen part is similar in shape to a given larger part of it.

accurate understanding of the nature of complex human information processing. The prototype of the *"Janusian framework"* is depicted in figure one above.

It provides a complex, four-way, interdependent, interactive model for thinking and acting that goes beyond the traditional linear processing associated with strategy, operations, and tactics and helps us understand what we could otherwise not decipher or comprehend. The framework provides the remarkable insight that the basic pattern of thinking and acting is fractal. In other words, conceptual patterns repeat endlessly, regardless of the field of study or social science in which we are interested.

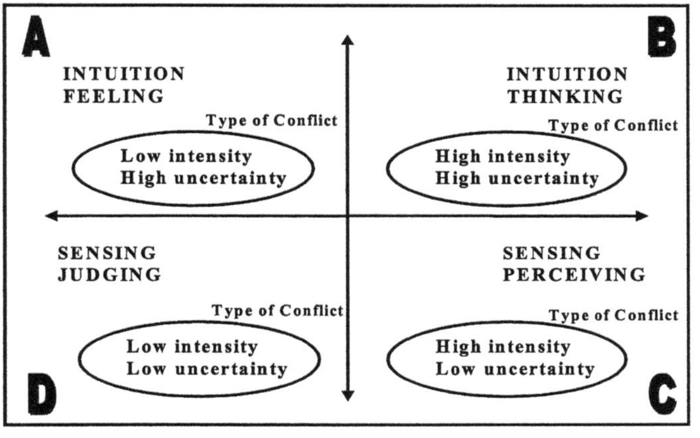

Fig. 2: The multidimensional paradigm

In 1965, Fred E. Emery and Eric L. Trist produced a seminal work describing *"the causal texture of organizational environments."* These descriptions were based on the degree of turbulence (placid, disturbed, and turbulent) and on the degree of interconnectedness present in the organization (random, clustered, reactive, and mutual). They organized these conditions into four environmental conditions: placid - clustered, turbulent - mutual, disturbed - reactive, and placid - random. The authors suggest there are corresponding coping mechanisms for each texture of the environment. The Janusian four-square (multidimensional) model applies to studying postmodern conflict like an overlay. It offers substantial insight into the war on terrorism. The authors propose that these conditions correspond to four types of conflict that also yield distinctive coping mechanisms.

The (Janusian) multidimensional paradigm corresponds remarkably well with Emery's and Trist's model. It is important to remember that the four environments depicted in the figure above always exist simultaneously. Strategy, operations, and tactics are relevant only in addressing type C conflict in its ideal form. Type C conflict does not occur in isolation from the other types of conflict but in combination with them; hence, strategic - operational - tactical thinking and acting is insufficient. Relying on strategy, operations, and tactics as patterned responses to conflict alone produces structural inertia. Unfortunately large hierarchical structured forces are often the source of undesirable structural inertia. The structural inertia that afflicts large organizations impedes their learning from small and dispersed operations.

Current Training Doctrines (TRADOC) cannot translate such experiences into force structure very well. When TRADOC tries to synthesize lessons learned, it tends to miss the essence and differences of each operation; hence, the value of change is diluted. Postmodern conflict should not only be examined through the constrained lens of strategic direction, campaigns, and tactics but also on a larger pattern of information processing through a pattern of multidimensional thinking. To embrace this new paradigm, we must transcend old ways of thinking and acting. For the military, a transformation in thinking and acting must accompany the armed forces' effort to transform its current organizational structure and equipment. In other words, military leaders must understand the fractal aspects of examining the approach to thinking about how and why permanent organizations are structured, as well as understand how to apply fractal notions to task-organized echelons. The flexibility that is achieved through task organizing must become common within the units themselves. Units must become more self-organizing. This framework permits all four types of conflict to occur simultaneously in various intensities.

Type A conflict: *Low-intensity, high-uncertainty.*[75] Tactics are insufficient, and strategy becomes important because survival becomes the dominant motivator. Strategy's purpose is to find the optimal location of safety in the environment.

Type B conflict: *High-intensity, high-uncertainty.* The turbulent field and the effects of unpredictable mutual causality shown in quadrant B are the dominant conflict types we are faced with today. In turbulent fields, boundary protection (strategy), linked tactics (operations), and order (tactics) no longer suffice. Events are so mutually causal that there is no longer a distinction between what was once considered tactical and that relatively autonomous and covert Special Forces teams or individuals sometimes make assessments and act without anything more than the strongly held values of their profession and country.

In reality, these ideal types of conflict do not occur apart from one another. The multidimensional framework permits all four to occur simultaneously in

[75] Paparone, Christopher R. (2001): 45-53 as well as the same, *"The Nature of Trust In and Between Organizations: What the U.S. Army Professional Needs to Know,"* unpublished.

remarkable patterns. The authors maintain that type B conflicts dominate the pattern of the present war on terrorism which has strategic significance. Today, the United States and other global players are deeply enmeshed in global and regional activities. Consequences that flow from activities in such an environment are highly uncertain and interconnected so that exclusively relying on the hierarchy of strategy, operations, and tactics becomes problematic because of the need to adjust rapidly to change.

Type C conflict: *High-intensity, low-uncertainty*. Coping requires an operational level of response because tactics and strategy are no longer sufficient. Global players see others emerge and occupy a similar status in our environmental domain. This type of environment dominated global conflict in the 20th century.

Type D conflict: *Low-intensity, low-uncertainty. "There is no distinction between tactics and strategy"* and the optimal strategy is just the simple tactic of attempting to do one's best to react in real time. At best, only partial stability of the environment can be achieved because strategic, operational, and tactical processes neither influence events rapidly enough nor are they flexible enough for the random nature of type B conflict. In B - D patterns of conflict stability can only be achieved through shared values because strategy, operations, and tactics cannot direct obligatory responses rapidly enough. Responses must come from a diffuse and flexible capability, which is the result of dispersed *thinking* and *acting*, based on appreciating the emerging situation and executing decisions based on a common set of values as well as habitual or professional action such as well – rehearsed standing operating procedures. For example, relatively autonomous and covert special forces teams or individuals sometimes make assessments and act without anything more than the strongly held values of their profession and country. U.S. Army Rangers displayed this idealistic *thinking* and *acting* when rescuing their fellow soldiers in Somalia in 1993. (Bowden, 1999) Strategy, operations, and tactics were so compressed that they were relegated to secondary criteria for decision making. In this situation, Rangers displayed an on-the-ground example of nonhierarchical *thinking* and acting that went beyond strategy, operations, and tactics. The leader's challenge becomes how to shape an appropriate value system that leads soldiers to do the right thing. A list of formal organizational values - something that the U.S. Army has recently developed may be insufficient.

The intuitive *"reasonable man"* view of humankind assumes dominance over the value- maximizing *"rational man"* concept. These autonomous social forces outweigh formal rules and structure, and free us to think in new configurations. Leaders can no longer rely on the dominant framework of processing information predominantly through the metaphysical interpretations of existentialism and rationality. Instead, leaders use the coping mechanism of shared values and specific routine actions to understand surprise or the enemy's intent. The difference we must perceive in our present condition is that, in light of the human processing the type B pattern of conflict requires, idealism must dominate multidimensional thinking and acting. In the postmodern conflict, we must understand not only where the enemy is located physically but also where he is morally - what drove him to accept a certain set of moral values that put him there in the first place. In the B - D pattern of conflict, we must provide at least secondary, but not necessarily less important, emphasis on the reactive coping mechanism of doing the best we can. The B - D pattern requires the simultaneous, instinctive actions of the highly trained tribal warrior and the cleverness of the 21st century soldier-entrepreneur.

The framework requires formulating new and complex recipes for thinking and acting in multiple patterns rather than embracing a singular one. Instead of using a linear thinking model to decide between competing values, the trick is to find a positive zone among them by using a nonlinear thinking model. All patterns in various mixtures are important to give relevance to the simultaneity of opposites, multiple time orientations, complex spatial relationships, and degrees of interconnected cause and effect. Implications for Military Leaders are shown in figure three below.

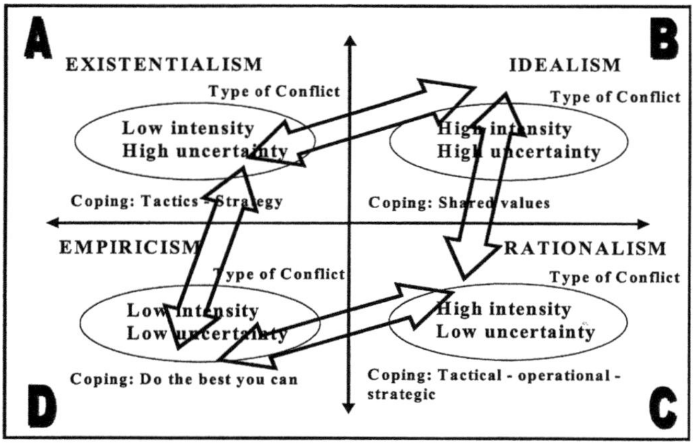

Fig. 3: *Thinking* and acting in opposites; embracing a unity of contradictions

Karl E. Weick (2002) explains that

> *"People try to fit novel interpretations and actions. New forms of organizing, such as the highly flexible network organizations, require a new power structure, something that the traditional military culture may find inconceivable: rank and hierarchical positional authority would have to give way to expert power and lateral forms of*

leadership into what they've known all along. And when something doesn't fit with the past, it's often discarded or misread."

The Emerging Dominance of Idealism

The postmodern era requires that our military leaders embrace new types of thinking and acting that affect postmodern living. These new types of thinking and acting exist simultaneously and with varying intensities and ambiguities as shown in figure three. When military leaders consciously adopt the Multidimensional paradigm as an intellectual framework, they add tremendous value and balance to their understanding of strategy, operation, and tactics. Military leaders must be able to link and integrate military, political, social, and psychological thought processes into a coherent approach to thinking about warfare and military operations in a complex world; otherwise, they might get locked into a linear mode of thought that not only limits their options but also leads them onto the wrong conflict path.

A greater emphasis on non-routine, appreciative inquiry and appreciative intelligence that considers multiple and different patterned sense-making across multiple time orientations, will be needed. How we work will also be affected, requiring multiple and simultaneous responses. In a B - D pattern of conflict, how we work will require an expanded reverence for philosophical interests such as clashes of the spiritual sensing of reality or value-to-value relationships.

Weick further explains that

> *"[Leadership] problems persist because [leaders] continue to believe that there are such things as unilateral causation, independent and dependent variables, origins, and terminations. ... Those assertions are wrong because each of them demonstrably also operates in the opposite direction: productivity affects leadership style, children socialize parents, responses affect stimuli, means affect ends, and actions affect desires. In every one of these examples, causation is circular, not linear."*

Leaders must look at today's conflict in terms of global sense making and realize that adversaries use values to motivate a different kind of soldier and to shape the battlefield. In today's conflicts, terrorists control the environment, a B quadrant activity, yet the United States has been responding bureaucratically, a type D response. Forming the Homeland Security Office, tightening airport and airplane security, and screening mail are all bureaucratic responses to the type B environment. To be successful in this war, western democracies must create an environment that is more turbulent and uncertain for the terrorists than the one they created for the West.

The terrorized Western democracies must seize the initiative in the B quadrant while sustaining the others. It must embrace dispersed, decentralized control and self-designing or self-managing capabilities that current systems do not promote or even allow to the needed degree. Many military commanders want to develop practical doctrinal prescriptions because they embrace the rational actor model. They should move away from such absolute linear thinking. Especially the US cannot afford to play a linear, tactical game of localized checkers while its adversaries play a patterned, global game of three-dimensional chess – one that uses a variety of moves employing different capabilities that can be sacrificed as long as the objective is not achieved.

Nonroutine, appreciative inquiry shapes the need to self-organized nontraditional intelligent organizations that consider multiple and different patterned meanings. Organizing requires emphasis on building teams, not monolithic, hierarchical units. In practical terms, this means organizing the military as an integral part of larger government capabilities based on multiple continua of adaptations needed for infinite configurations of conflict. There is a big difference between fighting a conflict by using current organizational capabilities and fighting a conflict by organizing around required capabilities. For example, in a future cold war with newly emerging global players, the US will be driven to forward deploy forces because forward deployed forces are within their capacity. Or do we adapt military, information, and economic capabilities to disrupt the adversary's self-governance fundamentally? One must ask if putting troops on the ground is the only solution available. But the military may be comfortable using traditional definitions of chaotic and complex situations; the real issue is whether the military will end up fighting with an erroneous conception of the pattern of conflict in which they really are. In essence they are fighting the wrong war.

The politico-hierarchical structure or *"polyarchy"* of the current western political-military system is inadequate for type B conflicts. New forms of organizing, such as highly flexible network organizations, require a new power structure, something that the traditional military culture may find inconceivable. Rank and hierarchical positional authority would have to give way to expert power and lateral forms of leadership. Instead of addressing levels of leadership – clearly a linear way of responding – the military must address patterned archetypes of the environment that require leadership effectiveness. Traditional top-down leadership can no longer be the only consideration for military management. Rather, the emerging pattern of conflict is best met with nontraditional, ad hoc flexible ways of organizing with members who are continuously self designing capabilities based on their unyielding shared values and mutual understanding of what is happening in the environment. This kind of adaptive organizing encourages innovative social and technical designs that are nonhierarchical and non-rational. Perhaps the special operations forces models become more attractive. Thinking and acting without orders must be acknowledged as sometimes appropriate in a government and within our own citizenry.

Today's environment will require developing and training confident, self-aware multidimensionally thinking and acting military and political leaders who are comfortable with lateral contributions regardless of the contributor's rank or position. This new paradigm requires a fundamental shift from the hierarchical *"strategic, organizational, and direct leadership"* espoused by current Western military doctrines to a new leadership paradigm which is all about role complexity.

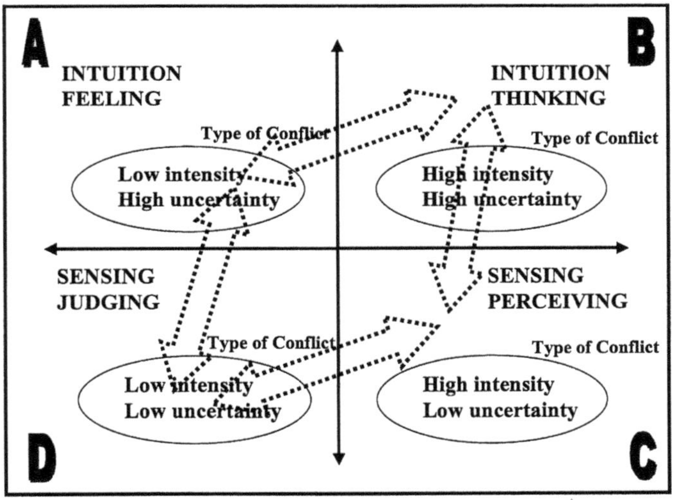

Fig. 4: Synthetic leadership paradigm

Those who sense and feel shifting organization and environmental patterns and adapt accordingly lead the most effective organizations. Modern western leadership has to transcend the need for hierarchical leadership because leaders focus on serving and developing high performance teams. The short names for these simultaneous roles are a) the *motivator*, leading commitment (type A conflict); b) the *vision-setter*, leading into an uncertain future (type B conflict); c) the *taskmaster*, leading against identifiable challenges (type C conflict); and d) the *analyzer*, leading compliance (type D conflict).

Embracing the notion that values associated with type B conflict and the values of habit associated with type D Conflict must become a key source of thinking and acting multidimensionally. This type of conflict cannot be coped with traditional tactical - operational - strategic linear ways of thinking. But military leaders should not throw away their learning styles associated with type A and C conflicts - they are still important and must exist simultaneously. However, western military theories and institutionalized thinking processes are still dominated by a paradigm of linear thinking; hence, the questions of strategy, operations, and tactics still dominate the paradigm for reasoning and responding. Unfortunately, western military teaching and training institutions are reluctant to make the intellectual pedagogical leap to include a more nonlinear, flexible way of theorizing and taking action.

Leaders must individually interpret the environment while maintaining others' visions of it. They must compete while cooperating, must obey orders, rules, and doctrine while simultaneously thinking and acting to defy them. Military leaders must be able to see themselves through the *"eyes"* of the supposed adversary to be fit for 21st century wars.

References

Bowdish, Randall G. (1999): U.S. Navy, Military Review December-February.

Bullis, Craig (2003): Developing the Professonal Army Officer, Implications for Organizational Leaders, Military Review May-June.

Everts, Steven (2001): Unilateral America, Lightweight Europe?: Managing Divergence in Transatlantic Foreign Policy. Centre for European Reform working paper, February.

Kagan, Robert (2003): Of Paradise and Power, New York, Knopf.

Paparone, Christopher R. (2001): U.S. Army Decisionmaking: Past, Present and Future. Military Review, July- August, 45-53.

Paparone, Christopher R.: The Nature of Trust In and Between Organizations: What the U.S. Army Professional Needs to Know. Unpublished, Figure 3, Comparison Chart for Janusian Conflict Types.

Strategic plan for Transforming DoD Training, March 2002.

United States Army White Paper, Concepts of the Objective Force.

Weick, Karl E. (2002): The sociol psychology of Organizing. New York, McGraw-Hill, 1979. In: Military Review Jan - Feb.

Yeakey, George W. (2002): Situational Leadership, Military Review January-February.

Wolfgang ROYL, University of the Bundeswehr, Munich

Moral Balance in the Face of Asymmetric Warfare

Conventional warfighting already stresses the moral awareness of both officers and soldiers to a more or less serious degree. This moral stress is exacerbated dramatically in asymmetric warfare.

Assuming that soldiers and officers alike have been brought up in their civilian lives to avoid violence and to distinguish between good and evil, then we may certainly consider wartime service as an extreme situation. The behavior acquired through training to ensure individual survival and the destruction of the enemy in combat demands a dismissal of civilian moral norms. The exigency to make such a dismissal becomes the more acute, the less the potential adversary respects the international law of war and the conventions regarding the treatment of wounded and prisoners-of-war as well as the use of outlawed weapons.

Terrorists are such adversaries. Thus, they are not entitled to the treatment to be accorded to lawful combatants. If, as a part of their asymmetric approach, they deliberately target innocent civilians, as they did on 11 September 2001, they abominably violate the notion of a just war. Lacking both the personnel and the material resources, these deliberate culprits, however, are unable to fight such a conflict by conventional means.

May such a terrorist rightfully claim an ethical legitimation if he follows his moral convictions and takes up the fight both against personnel and material targets that represent the industrial nations' claim to domination? Does he act for ethnic entities that have reason to feel oppressed and humiliated by the pre-eminence of the industrial nations? Does the history of colonialism reveal facts that may motivate a terrorist to sacrifice his life in deliberate asceticism to the claims for a just balance between rich and poor, between the godless and the faithful?

For a cautious answer to the question of how to balance, or better, to eliminate the antagonism of differing moral values, should be called attention to various international movements that have been attempting to aggregate individual feelings of justice at a nation-state and at an international level. If this idea gained acceptance, a world ethos would have to be observed in all political action. Hans Küng has presented an option that, by extension, is also supported by a well-known organization, the InterAction Council. With his Universal Declaration of Human Obligations he has challenged everyone to respect the moral force of human rights. All that remains at this point in view of the provisional character of these deliberations are reflections of whether a balance of power with a stronger executive branch were possible within a global government based on a democratic constitution; one that the United Nations Organization does not currently possess. This could evolve into a hegemon utterly convinced of the need of shared democratic living conditions for everyone that administers power under the rules of a universal constitution and restores rightfulness whenever it is threatened.

The Pervasiveness of Asymmetric Threats

Whereas previous wars were fought within the bounds of international conventions, asymmetric warfare is not governed by similar agreements. Here, the inferior contestant attacks like David against Goliath with unconventional means and methods. *"The figure of David enjoys a lasting popularity. It stands for the victory of the archaic over progress, for the victory of agility over stagnation, of guilefulness over vanity, of stratagem over brute strength."* (Sofsky, 1996: 28)

It would be a paradox indeed to expect underground fighters, who, by definition, plot in secrecy and vanish when they have delivered their blow, to take any interest in any binding agreements similar to the law of land warfare on the moral aspects of their manner of making war. The antagonistically stimulated moral values of the anonymous fighters for their faith will have to be accepted as the driving motive for their deliberate violence: take action to inflict the harshest punishment possible for all the world to see on the representatives of the capitalist world order that have denied us our cultural, in particular our economic equality.

Such a terrorist imperative relies on injustice in the distribution of potentially universal wealth and, from a global perspective, also on instances of such injustice in military history. It documents how leading industrial nations acquired great wealth through military conquests and subsequent colonial exploitation of the territories thus gained at the expense of the vanquished nations. It is no secret that the British maintained a feudal rule in India, the Belgians in the Congo, and the Dutch in Indochina for more than a century. In return, these nations offered their militaries a high standard of living both on regular service as well as afterwards. The gains from the economic exploitation of the resources found in the colonies and mandated territories permitted the construction of buildings of impressive splendor, such as in Brussels. Despite all the awe and admiration felt when you look at these buildings, one should never forget the often times utter disregard for human lives and rights in which these pompously displayed fortunes were amassed.

Such a retrospective is not meant to justify acts of terrorism that are now committed to injure or kill people who do not bear any personal responsibility whatsoever for the politics and the actions of past generations. In this sense the terrorist attack on September 11, 2001 struck innocents indeed. The terrorists victimized these innocents because, in their view, the U.S.A., with its profit-oriented economic power, dictates the debit and deposit balance of other nations. Colonial wars have evolved into trade wars. In the process, the issue of a world ethos and of a global power to supervise and enforce moral correctness has remained unresolved.

Instead, the world's strongest military power and her allies, but also some allies alone, are involved in military confrontations that appear to have been instigated by the actors of terrorist networks. In such a situation of global-historical importance, the top priority must be the containment of terrorist activities and the search for an effective response to the conditions dictated by asymmetric warfare. The dimension this threat has assumed was shown by Cordesman who listed 44 organizations that instigate asymmetric confrontations worldwide. (Cordesman, 2002)

Nevertheless, we should take the license to ask how the moral rules of such asymmetric warfare should be shaped, both from the attacker's and from the defender's perspective. This is closely linked with the question of how human rights should be reflected within this antagonism of attacker and victim. Any response, if designed asymmetrically, would conjure up the dialectics of an applicability or non-applicability of human rights. So far, the United States Government has chosen not to extend the protection granted by the Geneva Convention to captured members of the Taliban, but to detain them outside the continental United States in Guantanamo Bay on Cuba. The applicability of American legal norms and jurisdiction has thus been reduced. Viewed from this perspective, a paradigm begins to take shape in the treatment of personnel engaged in asymmetric warfare. Whoever endeavors to make war outside the bounds of the International Humanitarian Law and the Geneva and The Hague Conventions, whoever disregards the Human Rights Charter of the United Nations here and now cannot claim their protection if apprehended. In conventional wars, franc-tireurs or partisans who deliberately took part in the fight without being lawful combatants were shot on the spot. This has never deterred people acting out of their convictions from accepting this risk. Gerstein (1981), however, pleads that terrorists should not be placed outside the legal norms. The reason he gives is that there is a general consensus of right and wrong that should not be disregarded in judging terrorist activities. Such disregard would damage the moral foundations of any nation governed by the rule of law. This is a good reason indeed, even more so because we should not bar the way for a repentant terrorist towards becoming a law-abiding citizen again by canceling the existing relationship between the individual and the law in his case.

Whoever has fought as a terrorist or rebel and has stood his trial may later become an official representative of a state that in turn condemns acts of terrorism vehemently and helps seek out the perpetrators. Thus, former underground fighters that never denied their involvement in terrorist attacks in the British Mandate of Palestine were later to hold important posts in the government of the newly formed State of Israel. In these functions, they were respected and diplomatically recognized by the former hostile nations whose soldiers they had ambushed and killed.

Occasionally, the metamorphosis from the terrorist to the statesman demands a criterion for its assessment to facilitate its comprehension and the distinction from the perpetrators of subsequent acts of terrorism. Therefore, we dare take recourse to the diagnostic approach of a branch of knowledge utterly different from military pedagogy. Ludwig Klages, psychologist and philosopher, latterly of Kilchberg (Switzerland), used the term of *"standard of appearance"* in his book on character assessment based on handwriting (1943). This notion became necessary to judge samples of individual handwriting from which the factors of *"evenness"* and *"regularity"* were so absent that it was impossible to present a valid diagnosis. To this end, the ambiguity of these symptoms of handwriting had to be decided: Should the graphologically relevant features of this handwriting be used with a positive or a negative connotation in the diagnosis of the writer's character? To this end it became necessary first to assess the total impression of a sample of handwriting for its standard of appearance. Klages explained: *"The more the uniqueness of handwriting was above the time-dependent average, the more should its individual characteristics be judged positively, and negatively in the opposite case."* (Klages, 1943: 40)

A similarly discerning view of acts of terrorism was presented by Dugard:

> *"If another's motivation is personal gain or advancement of an "unjust" political cause, he becomes an international terrorist, a* hostis humani. *If, on the other hand, his object is to oppose colonialism, racism or alien domination, and to assert the principles of self-determination, then he is not a criminal, but a heroic figure, engaged in a just struggle against the Twentieth-Century infidel."*(Dugard, 1981: 87)

Those affected directly by an act of terrorism either as a victim or as a surviving relative of a victim will vehemently voice their outrage at the idea that any act of terrorism might ever be judged positively. Nevertheless, it is done, both by individuals and by nations that consider the perpetrators of such acts committed in certain asymmetric conflict as freedom fighters, not as terrorists. This truism was formulated by Quainton as follows: *"One person's terrorist is another's freedom fighter only because moral judgments about ends and means differ,"*

(Quainton, 1982: 40) and points to an identified moral antagonism. Such a play of words addresses the moral values of the industrialized nations and of their governments that dominate the world economy in relation to the moral values of those states and nations whose members feel oppressed and exploited. In an explanation of his thesis of the clash of cultures, Huntington (1997) speaks of so-called fault-line conflicts that arise where cultural as well as economic differences are branded as antagonistic, as inhumane by the opinion leaders of the nations involved.

By committing a crime against humanity, such as the destruction of the World Trade Centers on September 11, 2001, the perpetrators wanted to call attention to the crimes they felt the nations of the western sphere of prosperity had committed. The standard of appearance of acts of terrorism or of asymmetric warfare committed to achieve a more universal implementation of human rights must certainly be rated superior to and cannot be compared with the acts committed by bank robbers, blackmailers, and of organizations that make money by smuggling illegal immigrants across borders, as well as rebel groups and warlords that are primarily motivated by personal gain.

The actions of such individuals and gang leaders display a *"standard of appearance"* that justifies a strictly negative assessment of all individual characteristics of the antagonistic range of options. We must, however, make a clear distinction between these people and those of an ascetic attitude who employ quasi-military means to demand economic and humanitarian compensations in the culture of disadvantaged nations. Nevertheless, the victims of such attacks consider such actions criminal, although they have been committed in the name of a higher form of justice, for example, in the name of Allah. We should not fail to mention that similar actions were committed against infidels in the name of Christianity during the Crusades. G. E. Lessing (1779) voiced a belated indirect acknowledgement of this guilt through the parable of the ring in his dramatic poem *"Nathan, the Wise."* More than 200 years ago, he postulated a concept of reconciliation well in advance of Hans Küng (1997) who sought to combine a world ethos with world politics and world economy. Although the representatives of spiritual and political life argue for world peace both rationally as well as emphatically, and although the United Nations Organization has codified appropriate demands, it has proved possible to establish the idea of an obliging and uniting moral of peacefulness only partially and temporarily.

Karl Otto Hondrich (1992) pointed out that war, not peace, plays the role of a teacher of nations. Unlike religious convictions or philosophical views, the experience of war generates a much more effective recognition that a compromise must be reached between previously divergent political interests, and be it only through the victor's dictum. The same applies to terrorism, a derivative of conventional warfare, as a teacher of nations. This instructional competence of terror is currently obvious in Iraq, where the costs have been carefully weighed against the benefits between a greater military and financial commitment by the USA and the coalition of the willing and the recently favored option of consigning the responsibility for administration and for nation-building to an interim Iraqi national government.

> *"A sociological estrangement of the moral perspective of war is required to recognize the purpose of war as being collective, yet reluctant learning. Then, the question will no longer be: How much suffering does war inflict on mankind? Instead, the issue would be: "How does it, through the sufferings inflicted by war, change their ways of living together, their interests, values, balance of power, in brief, their collective identity?"* (Hondrich, 1992: 8-9)

This quotation is a good opportunity to go into greater depth of the multi-leveled perspective Hondrich mentioned. Analogously to the differentiation between the macro, meso, and micro levels, any analysis of the facts should distinguish between the interests of the individual, of the state, and of the community of states. So far, all attempts to establish inductive relationships across several levels have proved futile: the individual's love of peace does not find its exact equivalent at the aggregate level of the *"nation state,"* nor can it be transferred to the aggregate level of the *"international community of nation-states"*. Hondrich noted quite aptly on this finding: *"Any attempt to visualize a world without wars must answer the question of how the society of the world should be able to learn without it.*" (Hondrich, 1992: 9)

For the time being, Hondrich's question can be answered merely by saying that the nation-states have to continue to train soldiers for the purpose of protecting their population by military means and methods against unexpected threats. The soldier is to guarantee the security and the freedom of the nation to which he belongs. This moral claim states at the same time *"that the individual's life is not the highest good, that it may have to be sacrificed to collective survival: This is also one of the lessons of war"* (Hondrich, 1992:42)

All soldiers, male and female alike, are thus challenged to embed their individual moral views of non-violent behavior in the interest of a higher entity and the thus required fighting spirit.

The Soldier's Morale in Conventional Warfare

The engagement of hostile targets is an established part of the training programs of all soldiers, in particular of those belonging to the maneuver arms. As soon as these touch on the willingness to kill an adversary if necessary, a conflict arises with inter-human attitudes and moral convictions that have become an integral aspect of the ego-identity through upbringing and education long before the choice of the soldier's profession. Pursuant to this previous socialization, violence against people is justified only in self-defense and should be avoided wherever possible. In civilian every-day life, the use of violence both against people and animals is proscribed. Offenders will face pursuit by the police and trial by the appropriate authorities. The law-abiding citizen is well advised to avoid, wherever possible, such situations that may lead to violence in his more or less immediate vicinity. Yet if he witnesses violent confrontations, he may be legally obliged to help the victim in certain situations.

When deployed on operations, either low-intensity conflicts or regular wars, the soldier is exempt from the full applicability of the penal laws of his country. In contrast to his civilian life, he is now expected to be able and willing to engage hostile forces successfully. He accomplishes this by, among others, rendering the adversary unable to continue the fight. To this end, he must not only be willing, but also capable of killing an adversary *in extremis*. Thus, a theoretical emergency becomes an everyday occurrence in the heat of battle: Either him or me!

In conventional military conflict, combatants treat each other this way in battle. They are encouraged not to regard the adversary as a human being that can look back on a more or less ordered life in which most of them have not resorted to violence against people. He has been brought up more or less effectively to tame or even to repress his aggressions against people and objects. To prepare the soldier for combat operations, the military educational process is designed to establish a pattern of behavior that includes violence both against persons and objects up to the adversary's complete physical destruction.

Whereas an engineer on home leave will likely warn passers-by against using a derelict stairway or about a slippery trail, he will, on operations, attach demolition charges to a bridge or mine a road to inflict the heaviest possible casualties on the enemy in order to reduce the enemy's combat power. The soldier is required to thus obey the lawful orders issued by his commanders. Both the attacker and the defender are fully aware of the effects their actions may have. Either they become a victim of war or make the enemy a victim of war. In combat training, they learn about camouflage and deception and how to ambush an unwary opponent deliberately. Within this premise of the symmetry of military transactions, the soldiers on both sides of the frontline may take it for granted that they are expected to render each other *hors de combat* by whatever means and methods available. They know what they have to expect as soldiers. The infantryman who has to fight man against man may get a face-to-face impression of the effects of his weapons. In addition, he may also be an eyewitness of the effects on the enemy of artillery fire and air strikes in a combined arms battle. Whereas the frontline soldiers on both sides generally regard war as sheer madness and fight for their own survival, the personnel in the command posts further to the rear, at the meso level, act to implement strategic planning and the overriding political goals. The stronger battalions defeat an inferior opponent using the means and methods permitted by the international Law of Warfare and by the Hague Conventions. The starting situation is symmetric because both sides hope for victory, but this balance is disturbed and an asymmetry develops when one side begins to triumph over the other.

The Soldier's and the Terrorist's Morale in Asymmetric Warfare

Throughout its history, conventional warfare has generated horrors of such dimensions for soldiers and the civilian population alike, that it has regularly spawned hopes of a peaceful solution to clashes of interests between individual states and between military alliances. After both world wars as well as after the so-called *"Cold War"* between the NATO allies led by the U.S.A. and the eastern block under the leadership of the Soviet Union, the realization has gained ground internationally that war as a means of international enforcement of interests should be permitted only in exceptional cases.

The peaceful intentions of the League of Nations came to fruition and saw their implementation in the United Nations Organization (UN) that has remained exemplary despite repeated friction. In the UN Charter the member nations have expressed their determination to maintain peace. The organizational statute of the UN assigns the Security Council the task of taking the ultimate decision if military action is to be taken against states that pose an obvious threat to international peace and security. In the past, such a regimen has proved capable of preventing or containing impending major military confrontations in coordination with the General Assembly of the *"United Nations."* Although it leaves some room for improvement, this highly positive tool of achieving a consensus within the UN has now been challenged by quasi-military forces that do not claim combatant status for themselves and have used an exorbitant measure of violence against individuals and objects and continue to do so. The portent terrorists gave with the destruction of the World Trade Center and their successful attack on the Pentagon in Washington, D.C., on September 11, 2001, has bestowed an extraordinary measure of topicality and explosiveness to what had previously already been known as asymmetric warfare. The dimensions of this threat are great because the perpetrators of this terrorist attack have announced that others will follow. Thus, it is probably safe to assume that the attacks on the US troops in Iraq or the synagogues in Istanbul[76] are manifestations of the strategy of the Al Qaeda network. Using airliners as projectiles to bring down a symbol of capitalist world trade and yet to remain

[76] News report of 15 November 2003.

under cover has given the signal for worldwide efforts to take down this criminal organization. It appears apt to refer to this postmodern philosophical background in order to bring the linear expectation to an end that the physical destruction of the perpetrators alone would suffice to make the motives for the growing terrorism disappear.

Acts of terrorism are designed to destroy peaceful coexistence, both from the perpetrator's and from the victim's perspective. Suicide attackers have made an active decision to die in order to take others that have been declared enemies with them. The symmetry of killing, one life given for another, is distorted into an asymmetry when an assassin kills or wounds larger numbers of enemies. A defender also acts asymmetrically when he uses superior military means to inflict greater damage on the assassin's people than he himself sustained in the original attack. Acts the defender regards as a heinous crime are judged heroic achievements by the attacker's supporters. In the case of the Palestinian suicide attackers, such acts are rewarded materially by financial support to their families and spiritually by the promise of happiness in the afterlife. The question of whether these assassins act out of their own free will or out of religious or social compulsion is left unanswered here. What counts in this context is that people who sacrifice themselves for supra-individual goals act morally within the sense of their valid ethos of obligations. Legitimized by their own canon of military duties, the attacked defend themselves against terrorist attacks and prepare to deliver counterstrokes. The difference between combatants and the civilian population disappears in the process. Innocents on both sides come to grief. The international rules of warfare do not take effect in asymmetric conflict. Sanctions imposed by the UN can only be imposed effectively on one of the adversaries, for example, the Israelis, because the Palestinians are not organized into a state. This is also the reason why they do not have any armed forces for their defense. Due to this asymmetry in international law, the Palestinians who fight the Israelis by asymmetric means are not covered by the International Law of Land Warfare. As a consequence, this law is not applicable when suspected terrorists are liquidated and innocents or their property sustain injuries or damage in the process.

Regular soldiers involved in asymmetric operations, for example, by taking down the suspected shelter of terrorists without meeting any resistance or as pilots ordered to liquidate terrorists concealed in a refugee camp, may feel moral compunctions because they will necessarily cause so-called *"collateral damage."* The troops that do the fighting should not be left to cope with personal feelings of guilt alone. If asymmetric combat operations are expected with greater frequency, then we should start a moral discussion to justify the inevitable victim strategies, possibly combined with the introduction of rituals to absolve individuals from involuntarily committed crimes against humanity.

It will be necessary to distinguish such procedures very sharply and clearly from the rituals of justification, by which the National-Socialist strategists of the holocaust attempted to rid themselves and their agents of moral scruples that inhibited their *"performance"* between 1939 and 1945. The victims were declared parasites of the national community, or racially inferior forms of life that had to be eradicated. Anybody who did his best in this respect rendered sterling service to his *Führer* and his nation. This historical reminiscence may in no way be compared with the motivation that made Israeli fighter pilots refuse to carry out retaliatory attacks against Palestinian refugee camps. The soldier's moral dilemma assumes existential proportions if he feels responsible because his weapons would not only kill the actors of terrorism, but also innocent civilians. Arreguen-Toft rightly pointed out (2001) that asymmetric warfare may lead to barbarism on both sides. On a comparable occasion, Ditzer demanded:

> *"Pursuant to the principles of leadership that apply to the* Bundeswehr, *it must never happen that a soldier is systematically exposed to traumatizing experience without the means and methods at his disposal to restore the balance of his ego-identity."*(Ditzer, 1994: 47)

Moral Values on the One Side and on the Other

It was Lawrence Kohlberg (1973) who saw his task in life in assuming a universality of behavioral moral values and in proving this by intercultural comparison. The biography of this outstanding moral psychologist certainly allows us to see the motivation for his overriding interest in terminating the discrimination and persecution of minorities by public moral education. Such an education program was to be based on his model of the incremental development of moral awareness. Kohlberg's theory of the genesis of moral awareness states that both individual and collective behavior may evolve through education and personal experience from egocentric self-centeredness via respect for others to an internalization of justice as a guiding concept.

Kohlberg and his collaborators were not so unrealistic to expect an optimization of the moral behavior of their colleagues and students at Harvard University from a moral theory and the applied results of their research, much less a national moral movement. Instead, the systematic continuity of such teaching and research in the human sciences may be expected to cause the political class as well as the citizens ruled to increasingly approach those moral concepts that favor rational, peaceful attitudes and convictions instead of military solutions to all conflicts. The UN, the International Monetary Fund, and the International Court of Justice mark milestones on the way to world-wide peace and international cooperation. To safeguard such a development, we need military power that is capable of effectively countering unexpected threats and the associated military risks. Conventional warfare had always been costly both in terms of human lives and money, but the cost may even be higher in military solutions to asymmetric conflicts and in the neutralization of such conflict potentials. Huntington (1997) has illustrated his thesis of the clash of the cultures with a scenario in which Islamic nations launch an attack to liberate themselves from their oppressors, the western states, once and for all. The demographic developments in the Islamic states turns out to work against the demographically weak nations of the Western civilization. The determination to realize the claim to political equality and cultural recognition may be achieved by a continuous subversion of the Western welfare states, but also by a quasi-military liberation campaign with the use of force against weaknesses in the information system, the water supply system and against important establishments of trade and commerce.

What we consider a crime, a breach of International Law and of The Hague Convention of Land Warfare is hailed as an act of heroism by both the actors and their perpetrators. Thus, the moral balance is disturbed by terrorist activities by both sides within asymmetric warfare.

If the government representatives of Western nations read up on the history of their national campaigns of conquest and their practices of colonization, then they would have an opportunity of viewing a more self-critical perspective of the current war against terrorism and the propagated wars against rogue states in relation to their own history, including their own military history. This, of course, is not meant to say that subsequent crimes are cancelled out by previous crimes.

Cancellation of Antagonistic Morals before a World Ethos

An objective observer cannot help testifying that persons waging an asymmetric war for higher, in particular selfless reasons do act on a moral awareness. From the manner in which such awareness manifests and reveals itself it may claim a high standard or appearance and may thus require a positive judgment. As a consequence thereof, rebellion, terror, and resistance may, both from the underlying intentions and from the criteria by which they manifest themselves, pursue a partially valid moral concern. This statement applies when such actors or their national representatives are denied an opportunity to demand respect for their political and religious convictions through negotiations, this means in a non-violent manner.[77] (Münkler, 2002) A higher motivation has to fulfill the

[77] *"Thus the history of warfare since the middle of the 20th Century may be described as an increasing development towards asymmetric conflicts: The development of world-political asymmetries caused by the obviously unmatchable economic, military and cultural-industrial superiority of the USA has occurred alongside an asymmetrization of warfare*

requirement that they must be designed for a common good and thus transcend any hedonistic desires. The history of mankind is full of examples how insurgents, partisans, *franc-tireurs*, *guerilleros*, were supported and honored by their own ethnic group because they fought as freedom fighters and applied violence against oppressors. Such individuals or groups of individuals are wrongly labeled terrorists. Instead, they should be accorded the Stage 4 of Kohlberg's scale of moral awareness. An individual who has attained this level through upbringing and education respects government authority, law and his duties to preserve an existing social order, no matter if this is defined in secular or religious terms.[78](Kuhmerker, 1973)

In view of the antagonistic positions held, for example, by the Palestinians and the Israelis, both parties feel from such a moral awareness that they are in the right and, therefore, justified in resisting their political opponent. Israel as a military power has an advantage at both the strategic and the tactical levels over the Palestinians that are only provisionally represented by an autonomy authority. This results in an asymmetry with respect to military capabilities and to the status under International Law at the expense of the Palestinians. This asymmetry remains fixed because no other Islamic nation in the Middle East has the conventional warfare capabilities required to enforce Palestinian interests. From a western perspective, such enforcement would be a great tragedy. Leaving western preferences aside and looking objectively only at the features we can discern in the Middle Eastern theater of operations. Then we also recognize only the acts of terrorism, or acts of war, of militant Palestinian organizations, such as Hamas, at an operational level also the considerably higher birth rate in the Palestinian autonomy areas and the worldwide public condemnation of Israeli retaliatory strikes.

The threat to Israel grows demographically because the social position of women within certain interpretations of Islam prevents birth control and renders female aspirations at education both impracticable and undesirable. Compared with the relatively more restrained mourning the Israelis show in public when friends or family members have been claimed by terrorist attacks, the adherents of Islamic faith fully release their feelings in plain view of the international media when they bury their martyrs. Therefore, we would be well advised to remember the measure in which the public presentation of Palestinian martyrs, in particular after their sacrifice, the often religiously motivated ban on female educational opportunities, the male domination under Islamic law and the

---

through a relocation of the combat zones, a redefinition of the means of warfare and a mobilization of new resources." (Münkler, 2002: 53)

[78] Stage 4: *"An orientation toward authority, law, duty, the maintaining of a fixed order whether that is secular or religious."*

resulting higher birth rate all combine to bring about a creeping reversal of the political balance of power in parts of the Middle East.

The actors on both sides of a historic fault line draw the arguments for their conduct of the conflict by citing partial moral reasons. Both the Palestinians and the Israelis were promised material resources in terms of land and water by the former mandatories. In their struggle for these resources both sides violate the moral core commandment: *"You shall not murder!"* both directly and indirectly.

We owe Hans Küng (1997), among others, for a complex diagnosis of the political worldwide situation. The states that had hitherto pursued *realpolitik* were unable to cope with the international conflicts. The desire of the nation-states to preserve their power has so far prevented any real compromise between the interests of the industrial nations versus those of the threshold and the developing nations.

In order to prevent world-wide chaos, Küng pleads for an agreement among the principal religions on a world ethos and on the moral core elements of an orderly coexistence. He challenges the churches to relinquish their power like the nation-states following a *realpolitik*. He believes this is a reasonable price to pay. The representatives of the world religions should set a positive example to the politicians in government office in achieving a consensus with their various denominations on their commonalities.

> *"There will be no peace among the civilizations without peace among the religions! And there will be no peace among the religions without a dialogue among the religions."* (Küng, 1997: 131)

Such a proposal is not as utopian as it may seem. After all, a *"Parliament of the World Religions"* was constituted successfully and a declaration on world ethos was negotiated from August 28 through September 4, 1993 in Chicago. Remarkably, yet also typically, the former heads of government and minister presidents adopted this declaration on a world ethos during the meeting of the InterAction Council in Vancouver on May 22, 1996. This declaration lists, and this is worthy of note, not only the human *rights*, but also the human *duties* in codified form. These statements were made in the *"shall"*-form that is so typical of moral commandments. Article 6 of this declaration reads:

> *"Controversies between states, groups or individuals shall be settled without recourse to force. No government shall tolerate or contribute to acts of genocide or of terrorism, nor shall they misuse women, children or any other civilian individuals as a means of waging war. Every citizen and every bearer of public office shall act in a peaceful and non-violent manner."*

The InterAction Council celebrated its 20th anniversary in Berlin on June 5, 2002.

In his welcome address at the NATO Summit in Prague on November 21, 2002, German Chancellor Schröder mentioned, among others, the common threat posed by international terrorism. The declaration at the Prague Summit by the heads of state or of government of the NATO countries contains the following as its 7th point:

> *"We welcome the resolve of Partners to undertake all efforts to combat terrorism, including through the Partnership Action Plan against Terrorism. We will also continue to further enhance interoperability and defense-related activities, which constitute the core of our partnership."*

In a NATO declaration, there is no room for self-conscious statements on the history of terrorism and its causes or for confessions of guilt for causing terrorism and its manifestations in asymmetric warfare. Thus, to complement the required military defensive measures against terrorism, it becomes only the more important that both the *"Parliament of World Religions"* and the *"InterAction Council"* have highlighted vehemently the need for patterns of conduct that address all people.

If we take the blessings of the activities of non-governmental organizations, such as the International Committee of the Red Cross, into consideration, we may conclude that a public moral awareness is increasingly taking shape. Let us not forget the scientific foundations laid by Lawrence Kohlberg and his students in their comprehensive theoretical and practical works. They, too, strove to prove and to promote the universal validity of the developmental stages of individual moral thought. The topic discussed at ethics congresses, conferences on human obligations, and at NATO summits require a moral awareness as Kohlberg has defined it as the sixth stage of his empirically verified tenets of the incremental development of moral awareness: The individual orients himself by universally valid ethic principles and his practical actions by the generally recognized principles of justice, the respect of reciprocity (*"The Golden Rule"*), and the same human rights for all. It is probably not without emotion that you will hear (read) what Kohlberg reported about a ten-year-old student using his own words to explain reversibility. Treating others as you would like to be treated:

> *"Well, it's like your brain has to leave your head and go into the other guy's head and then come back into your head, but you still see it like it was in the other guy's head and then you decide that way."*(Kohlberg, 1973: 642)

It may be assumed that the bearers of political mandates, which are responsible for the preservation of peace or for the employment of the military, are essentially capable of similar reflections to those of this ten-year-old. The diplomatic reservation of the transfer of individual behavior to national and international politics is still strange to this boy. Küng and his collaborators are convinced that moral antagonisms may be eliminated in a dialectic sense through an acknowledgement of a world ethos.

References

Arreguin-Toft, Ivan (2001): How the Weak win Wars? A Theory of Asymmetric Conflict. In: Int. Security, Vol. 26,1, 93-128.

Cordesman, Anthony, H. (2002): Terrorism, Asymmetric Warfare, and Weapons of Mass Destruction. Westport, Con., Praeger.

Ditzer, K.H. (1994): Neue Aufgaben und soldatisches Selbstverständnis. Bonn, Katholisches Militärbischofsamt.

Dugard, John (1981): International Terrorism and the Just War. In: Rapoport, D.C.; Alexander, Y. (ed.) The Morality of Terrorism. New York, Pergamon Press, 77-98.

Gerstein, Robert S. (1981): Do Terrorists Have Rights ? In: Rapoport, David, C. & Alexander, Yonah (eds.): The Morality of Terrorism, New York, Pergamon Press, 290-307.

Hondrich, Karl, Otto (1992): Lehrmeister Krieg. Reinbek, Rowohlt

Huntington, Samuel, P.(1997[3]): Der Kampf der Kulturen. München, Europaverlag.

InterAction Council (1996): Die Allgemeine Erklärung der Menschenpflichten des InterAction Council. http:/uranus.ecce-terram.de/zeit-archiv/1997/41/prambel.text.

Klages, Ludwig (1943[22]): Handschrift und Charakter. Leipzig, Ambrosius Barth.

Kohlberg, Lawrence (1973): The Claim to Moral Adequacy of a Highest Stage of Moral Judgment. In: The Journal of Philosophy, Vol. 70, 630-646.

Küng, Hans (1997): Weltethos für Weltpolitik und Weltwirtschaft. Munich, Piper.

Kuhmerker, Lisa (1973): Growth Toward Principled Behavior: Lawrence Kohlberg's Studies of Moral Development. J. Moral Education, Vol. 2, No. 3, 255-262.

Lessing, Gotthold, E. (1983): Nathan der Weise (1779) Stuttgart, Philipp Reclam jr.

Münkler, Herfried (2002): Die neuen Kriege. Reinbek, Rowohlt.

NATO (2002): Prager Gipfelerklärung der Staats- und Regierungschefs vom 21.11.2002.

Quainton, Anthony, C.E. (1982): Moral and Ethical Considerations in Defining a Counter-Terrorist Policy. In: Rapoport & Alexander, op. cit., 39-45.

Rapoport, David, D. & Alexander, Yonah (eds.) (1982): The Rationalization of Terrorism. Frederick, MD, Univ. Publ. of America.

Royl, Wolfgang (2004): Moral und nicht-konventionelle Kriegführung. In: Schröfl & Pankratz: Asymmetrische Kriegsführung. Baden-Baden, Nomos. 329-344.

Sofsky, Wolfgang (1996): Traktat über die Gewalt. Frankfurt/Main, Fischer.

Edwin R. MICEWSKI, National Defense Academy, Vienna

ETHICAL, POLITICAL, AND MILITARY IMPLICATIONS OF ASYMMETRIC CONFLICT AND WARFARE

Introduction

Reflecting on the theme of military ethics makes one believe – very much in line with the 19th Century German philosopher Arthur Schopenhauer – that *"human existence, in the end, is justifiable as an ethical phenomenon only."* (Schopenhauer, 1995: 200)

Given this ethical character of human existence, it can be inferred that matters of war and peace pose perhaps the most fundamental challenges to any well-reasoned concept of (political) right and wrong. Accordingly, the moral responsibilities on the part of military leaders are immense, especially in the light of the contemporary, significantly altered face of armed conflict and war. The following essay therefore attempts to link thoughts on Just War – or rather, ideas on how to morally justify the use of violence and force – to the contemporary conditions of war and armed conflict. These are conditions that bring the question of human dignity to the fore more than most of the other issues that challenge us morally in our day.

Due to the growing influence of non-state and sub-national actors who pose their claims violently, the conditions of armed conflict and warfare are – in the parlance of today's political philosophy – mostly described as asymmetric. Thus, my concern here is not - militarily speaking - the asymmetry on an operational or tactical level where good leadership is always about seeking a temporary asymmetric advantage in order to prevail militarily. The topic here is this unprecedented change in international affairs, the small-scale eruption of new forms of warfare with different players who do battle without state-supported armies and resources; carried out by political and ideological, ethnic or religious factions arguing their case through violent means, be it in civil or tribal wars or terrorism acting on a regional or global scale. In other words: this essay refers to asymmetry as a political one, originating from a non-state subject that engages in violent conflict, and looks at how they might justify their actions.

To set the remarks in a broader context, we will start out with a brief look at the etymology of the term asymmetry. This will be followed by some basic philosophical and historical thoughts on war and its development, mostly for the reason that the political and ethical nature of asymmetric conflict and warfare cannot be grasped without understanding the nature of symmetric armed conflict - under which conditions the theory of Just War was born. Then, the attempts to domesticate war will be described and some important elements related to the Just War Theory beyond its principles will be outlined. Next will be an examination of the political asymmetry in warfare of our day – of which 9/11 was the most striking example – combined with some normative ideas for an ethical evaluation and decision-making for the use of violence and asymmetric force, this predominant new phenomenon of international politics. The conclusive part will talk about the transformation from an instrumental to an existential understanding of war and will suggest some imperatives for the future.

On Symmetry, Asymmetry, War, and Just War

Etymologically, the term asymmetry stems from the ancient Greek *asymmetria* and has the meaning of the lack of an appropriate proportion, a proper relation, and the absence of a common gauge or measurement. Asymmetry delineates an irregularity that disturbs harmony as the foundation of the true, the good, and the beautiful and can never serve as a prototype, an ideal paradigm.

However, in the 3rd century B.C., Archimedes founded a theory that doubted the ideals of symmetric perfection. His deviations from symmetry and his asymmetric scientific and technological inventions determined the first asymmetric war of known history, the Second Punic War between Rome and Cartage. (Mongini, 2003: 15)

War has been defined by Clausewitz as *"Being nothing but a duel on a larger scale"* and a *"Trial of moral and physical forces decided through the medium of the latter."* (Clausewitz, 1976: 75 and 127) As such, war appears to be something as eternal as peace. Both are associated in a dialectic relation to one another and seem to alternate continuously. It appears as if the ontological principle of our existence is truly contradictory, or rather, conflictory, meaning that the potentiality of evil, badness, wrongdoing, crime, and war, as the most severe manifestation of human struggle, might be an inexorable ingredient of human existence.

Nevertheless – in this conflictory, conflict-ridden pattern – war with all its strain, suffering, and exhaustion is holding the place of the *a priori* negative and unwanted, as humans normally strive for peace. Peace is the maxim and the question is how to maintain peace or, if this is not feasible, how to restore it as quickly as possible.

It is for this reason that thinkers have always ventured to grasp the nature of war, and how it could be prevented, but – if wars occurred – to find ways how to wage them efficiently, meaning how to end them as quickly as possible and do so with the least damage and loss of resources.

This introduces two aspects that should be mentioned particularly because asymmetric warfare precisely operates with the destruction of these defining elements of traditional warfare – victory and time. Victory in war is a matter of depriving the adversary of his instruments of power and freedom of action, or, to use Clausewitz' words, to break his (political) will. Victory is no end in itself; it serves a political aim, wherefore war is a *"serious means for a serious objective."* (ibid: 86) The idea to end war efficiently and restore peace necessarily implies the aspect of time. In order to maintain the existence and full authority of a political community, war has to pay off in the sense of achieving the political aim through victory as economically as possible. A protracted use of resources certainly does not serve this goal.

While the reasons to go to war and the face of war have changed constantly over the centuries, the goal of warlike activity remained principally unchanged: to be victorious. Sun Tsu made this clear already some 2500 years ago: *"In war, then, let your great object be victory, not lengthy campaigns."* (Sun Tsu, 2000: 18)

In the twentieth century a prominent representative of revolutionary theory of war and a theoretician of guerilla war (as one form of asymmetric warfare), Mao Tse-Tung, embraces the principle: *"War as the highest form of struggle between nations, states, classes or political groupings ... is applied in order to achieve victory."* (Mao Tse-Tung, 1966: 45)

It therefore comes as no surprise that the politician and military leader, Archduke Charles of Austria (who was one of the very few to defeat Napoleon I in battle), commences his *"Principles of the Higher Art of War for Generals of the Austrian Imperial Forces"* by saying:

> *"Since war is the utmost evil that can happen to a state, a nation, the main concern of the Emperor, a commanding General, has to be the mobilization and application of all means available right at the beginning of war in order to keep war short, and to decide it in the most favorable way possible."* (Karl von Österreich, 1974: 2)

These considerations - the aspects of victory and of time - dictated the rationale of strategic conception especially since the Westphalian Peace Treaty in 1648, which put an end to arbitrary warlike conflicts between religious and ethnic groups, clans, and factions during the Thirty-Years War. War became nationalized and politicized. It became an expression and a legitimate means of political sovereignty, with the borderlines of military conflicts basically corresponding with (nation-) state borders.

Military theory and strategy have been guided by the Clausewitzian assumption that war is a rational undertaking, a reflection of national interest, and an extension/continuation of politics by other means.

This development resulted in the fact that certain symmetry in strategic thinking, war-preparing efforts, and the configuration of military instruments emerged. The military instruments sent into wars and battle resembled each other significantly in terms of structure, organization, weaponry, etc. What distinguished them were their motivation, their quality in respect of military education and training, and the excellence and innovative power of their leaders.

Given the very nature of war it did not come as a surprise that the political and military symmetries were reinforced by the development of humanitarian standards, and moral and legal norms that were to be applied in armed conflicts and intended to restrict warfare and armed competition.

Those norms frame the Just War Theory as it emerged in the Western world in a century-long process of philosophical and theological thinking (St. Augustine, Ambrosius, Thomas Aquinas or the Jesuits from the School of Salamanca), finally being transformed (by Hugo Grotius) into the first attempt to establish International Law, binding and relevant for all civilized nations.

The approach constitutes something between pacifism and militarism. Whereas pacifists regard warfare as an untouchable evil, militarists regard battle as a positive good. The Just War theory represents a sort of conditional justification, which proves that it is possible to reject pacifism without embracing militarism.

The political ethics of Just War operates on two levels, the *Jus ad bellum* (right *to* war) and the *Jus in bello* (right *in* war) dimension. (Christopher, 1994) The first level aims at reducing or eliminating the occurrence of war, the second, where this proves impossible, is directed toward mitigating the consequences of war.

The first set of principles, the *Jus ad bellum* criteria, concerns the conditions that make the use of force permissible. Among them the most prominent are: a) War must be declared by a lawful authority; b) There must be a just cause c) War must be prosecuted for rightful intentions; d) War must be a last resort; e) There must be reasonable hope for success; f) The political objectives must be proportional to the costs of war (end-means relations); and g) War must be publicly declared (in terms of causes and contents). Naturally, under the primacy of policy in the Clausewitzian sense of war, the Jus ad bellum principles primarily constitute a political responsibility.

The second set of principles, the *Jus in bello* (Right in war) criteria, concerns the rules governing how war should be conducted, thus constituting a largely military responsibility. The Jus in bellum principles basically come down to a) War must be fought justly; and b) Discrimination must be observed (in the sense of noncombatant immunity).

In our context of human dignity of particular interest: a most decisive dimension of the Just War Theory is that the opponent is being regarded as a *Justus Hostis* (just enemy). In this period of the domestication of war the adversary was of the same (human) quality as oneself, subject to rights and duties of combatants as clearly outlined in the laws of war of international law. This mutual acknowledgement between sovereign states allowed for the instrumental waging of wars! The war was basically *"aristocratic,"* (Stadler, 2001: 24) with limited and clearly defined groups of people (combatants) fighting each other.

However, this changed along with the democratic revolution since the face of war always transforms in accordance with the political order within which it is being waged. The age of democratic wars brought about not only the era of mass armies, but also the end of the domestication of war.

Since victory in war has always been a matter of breaking the (political) will, the object of war changed to the 'demos,' the people, as they have become the bearer of the politically relevant will.

As an expected consequence, war turned its will-breaking destructivity toward the people, no longer limiting itself to the decisive encounter of combatant troops on the battlefield. A development that culminated in two world wars in the twentieth century with both wars providing for sufficient testimony (Dresden, Hiroshima) that the people became – although noncombatants according to the Just War Doctrine – a prime target of modern warfare. Subsequent to World War II it seems to be unacceptable to leave the authority of the *Jus ad bellum* to sovereign nation-states.

No wonder, therefore, that the UN Charter finally became an attempt to proscribe war as a means of putting through political objectives, which meant, at the same time, that the idea of a Just War – at least in its *Jus ad bellum* respect – was abolished from international law or, rather, limited to clear cases of self-defense. The justified use of force was limited to the idea of individual and collective self-defense according to Article 51 of the United Nations Charter.

The national logic to prepare for war – meaning for the cases of collective defense – was based on the idea to equalize potential asymmetries in order to maintain political stability.

Among the great powers this led to what has been called the arms race, for others it meant to integrate into alliances or to find asymmetric strategies (such as the area defense concepts of neutral states) that allowed for compensating the superiority of an adversary in order to deter him from commencing a campaign in the first place.

The Privatization of War – Political Asymmetry and its Ethical Evaluation

With the end of the bipolar world, global politics again entered into a new phase. The increasing de-politicization and growing influence of nonpolitical actors is once more accompanied by a profound change in the nature of armed conflict and war.

This process is undoubtedly promoted and exacerbated by two facts: First, that both the United Nations Charter and International Law are still almost completely oriented toward classical interstate conflict; and second, that the United Nations could not manage to monopolize violence and force on an international level - due to the decision-making process in the Security Council as well as due to a lack of means to enforce International Law.

War and violence, however, remain forms of political manifestations, nourished by aspirations of asymmetric forces to cause superior actors (organized political communities) to give in to the claims of perpetrators using asymmetric strategies. A dramatic inequality of military power is contradicted by asymmetric procedures of how to wage war, and a potentially unlimited logic of escalation. As a consequence, asymmetric warfare is not about victory or defeat, but rather about destruction.

To this end, the primary strategic significance from a military viewpoint is the dissolution of spatial and temporal limitations of traditional war. There is no front and rear anymore and there is no limited duration of battle and combat activity either. Nonetheless, what makes matters worse is the fact that the procedural and technological-material asymmetric strategies are combined with moral-legal asymmetries.

Ethical Evaluation

Most interestingly, the ethical challenge moves particularly to the fore when it comes to asymmetric policies since sub-national actors operate in a juridical-free or, as Kant has called it, moral-juridical space, in which both national and international statutory laws are being neglected – simply because there can be no statutory law that allows for violating the national or international monopoly of force. If asymmetric violence intends to justify itself, the justification can only be a moral-ethical one. (Micewski, 2003: 31-40)

With the exception of last-judgment-day sectarians or other irrational forces who are driven by apocalyptic visions and simply have a desire to destroy and demolish,[79] terrorists or other movements who pursue a rational cause have an interest in justifying their case – merely for reasons of publicity, which they need as a precondition for public support.

If any entity wants to justify actions that are ethically relevant, three typically distinct elements need to be involved: goals or intentions, means, and results.

[79] As, for example, the Sarin nerve gas attack of the Aum sect on the Tokyo subway on March 20th, 1995, with 3800 victims altogether, among them 11 dead.

To decide within this ethical triangle, for a trade-off between these elements, we basically find two dominant strategies: the *ends-based* and the *rule-based* approach. (Amstutz, 1999: 27-41)

The ends-based approach is generally defined as consequentialism or teleological[80] ethics and assumes that the morality of an action must be ultimately judged by the good results that are realized.

Consequentialism believes that the most important moral criterion is the overall outcome. The consequences of human choices are given priority; in other words, the moral legitimacy of an action ultimately depends on its consequences. Thus, the goals and means of political action must be morally justified by the results of such actions.

End-based thinking is an influential ethical strategy in domestic and international politics in great part because government decisions are judged generally in terms of results rather than intentions. As a result, political ethics tend to be outcome-oriented.

Given the indeterminacy of public policies, the moral assessment of political action from an ends-based strategy is thus likely to involve significant uncertainties and ambiguities. An even more important limitation is the inability of ends-based thinking to provide an authoritative ethical standard by which to evaluate action.

In short, although an ends-based perspective needs to be incorporated in political ethics, consequentialism alone does not provide a fully satisfactory ethical methodology for applying morality in international affairs.

In contrast, *Rule-based* analysis, known as deontological[81] thinking, denies what the consequentialist's analysis asserts, namely that an action is good or bad, depending on the goodness or badness of its consequences. Even if the action results in good outcome, the use of evil means (e.g. intentionally killing civilians) cannot be justified by good intentions (e.g. taking an enemy position).

Deontologists assert that actions should be judged by their inherent rightness and validity, not by the goodness and badness of policy outcomes. As a result, ethical decision-making is determined mainly by the morality of goals and intentions.

It is important to note that deontological moral obligations should be fulfilled not because they are more effective in creating a better, more just world, but because they represent what moral actions require. An action done out of respect

---

[80] From the ancient Greek word *telos*, meaning *end.*

[81] From the Greek word *to deon*, meaning *duty* or *obligation.*

for the moral duty has its moral worth not in the purpose to be attained by it, but in the maxim in accordance with which it is to be decided upon.

This Kantian-based perspective is important because it guards against the relativism of consequentialism. But how could anybody know what his moral duties are? Kant argues that the rightness of a rule or action should be based on the categorical imperative. This principle has two key dimensions: First, persons should be treated as having value themselves, and second, individuals should act in accordance with principles or maxims that can be universalized.[82] (Kant, 1993: 30 and 36) Thus, ethical decision-making should be judged on the basis of the extent to which a principle should be applied to others.

But now the questions arise: How can we make moral decision-making work in this three-dimensional context? How can we combine consequentialist and deontological claims? What is an appropriate approach to make the necessary trade-offs between means and results, translating ethical intentions and purposes into actions that will maximize desirable outcomes without violating moral principles?

Especially when it comes to justifying the use of forceful means, it is appropriate to follow the tradition of the prudent and wise practice as outlined by Aristotle and Thomas Aquinas. They considered it to be a matter of fact, as Aristotle put it, that

> *"The work of man is achieved only in accordance with practical wisdom as well as with moral virtue; for virtue makes us aim at the right mark, and practical wisdom, prudence, makes us take the right means."* (Aristotle, 1998: 155)

The prudence tradition has two distinctive features: First, it acknowledges the evil of human affairs and insists on the definitive authority of morality over circumstances, of "ought" over "is." Second, the tradition regards human virtue and personal character to be the crucial factors of morality, rather than religion or political ideology. The moral quality of leadership thus hinges upon character and personal integrity and not on technical proficiency. (Amstutz, ibid: 43)

When we act we want to bring about a desired result. In an ethical context, this desired result might be the warding off of an attack, saving a person's life, righting a wrong of whatever kind, eliminating or bringing about certain

[82] The first dimension corresponds with Kant's third formula of the categorical imperative: *"Act in such a way that you treat humanity, whether in your own person or in the person of another, always at the same time as an end and never simply as a means;"* the second is the well-know first formula of Kant's moral imperative which defines the universalizisability norm literally as follows: *"Act only according to that maxim through whereby you can at the same time will that it should become a universal law."*

conditions, etc. The desired result of my acting must translate into an intention or motivation – and this incentive must be ethically justifiable.

The context of result, intention, and means becomes particularly conspicuous and delicate in situations where physical violence or force is applied. Force and violence have in common that they involve the infliction of physical pain and injury. In the abstract, *violence* can be defined as the forceful encroachment upon the sphere of an individual's legitimate development of his free will. This makes it easy to conceive why the question of right and wrong is determined from the perspective of the victim who must sustain a restraining and debilitating effect on his organism and thus its free self-development. This direct idea of violence comprises a clear relation between a perpetrator and a victim, with the perpetrator directly inflicting damage on the victim. The damage primarily violates the victim's organism or its free self-determination. From this it becomes quite clear why, for instance, the incentive to satisfy a personal pleasure at the expense of one other's health or wealth will certainly not fall under the category of ethically justifiable motivations to act.

Now, when it comes to violence which always is a situation of social interaction, as in any situation of moral relevance, a dilemma situation of colliding interests emerges that has to be resolved by any individual (entity).

Ever since Greek and Roman antiquity philosophers have tried to find an answer to this question of conflicting interests. Particularly Aristotle posited that in the context of *violentia* – when it is about the justified use of force in a relation between a perpetrator and a victim – we encounter a situation of colliding legal goods in which it has to be decided *"which of two things to select, and which one of two maladies to bear."* (Aristotle, 1998: 49 [1110a])

Hegel differentiated that further by making clear that in a situation of distress, in which forceful behavior becomes inevitable, the preservation of qualitatively higher or quantitatively more comprehensive rights is justified even at the expense of the curtailment of the rights of others. (Lenckner, 1965: 84) In other words: The wrongdoer reduces himself, diminishes his human dignity, lessens his claim to enjoy human rights, and while he by his own doing justifies a moral hierarchy, he remains a human being.

Based upon this normative model it becomes obvious that the intentional putting through of right and justice can be justified as a last resort even by way of forceful means. The principle of last resort has to be understood factitiously and not pertaining to a succession of first peaceful and only then forceful measures in time.

The desired outcome – righting a wrong, maintaining or restoring justice – may only rest on the motivation to use violence and force when the use of force is unavoidably necessary. It is clear to any unprejudiced mind that violence and physical force may only be applied when it is indispensably necessary, when the act of violence is inevitable, and is the last resort to resolve a situation. So we encounter necessity and inevitability as normative ethical factors for possible justification of violence. Violence must be limited by purpose.

Cause and motive together must make the use of violent means inevitable – constituting the conditions for a last resort – in order to bring about a desired result of righting a wrong or eliminating certain conditions.

Transformation into Political Philosophy

If we want to translate this context of result and intention into an adequate political philosophy for the use of violence and force, it appears to be inevitable to introduce the idea of a moral order of priority in order to find moral orientation. This order should at least comprise of two levels.

The first level refers to what we might call primary or *transcendental* rights. Totally in accordance with the Kantian understanding of transcendental being "the condition of the possibility of experience itself" transcendental (primary) rights are those, which build the precondition to enjoy other rights. Hence, transcendental rights are the immediate human rights, such as the right to live and to enjoy physical integrity.

The programmatic (secondary) rights such as political self-determination, creation or maintenance of democratic living conditions or the just distribution of goods follow the transcendental rights and are based upon these in a normative ethical order of priority.[83] (Kersting, 1998: 36ff)

However, the specific challenge rests where this order is being turned upside down when, for the sake of putting through secondary rights such as the implementation of a desired social order, transcendental rights are being denied.

But this is exactly what we normally come across in most cases of asymmetric violence. Normally – at least not before he is using violent means – the asymmetric fighter is not threatened in his transcendental rights. However, he does not observe the fundamental, primary, transcendental rights of others, neither taking them into consideration as a physical cause nor as a moderating motive for his actions. In reference to terrorism it is precisely at this juncture that we become aware of the fact that terrorism is devoid of any serious ethics. Terrorism does not take these rights into account (or rather intentionally or unintentionally disregards these normative principles), nor do terrorists hesitate

[83] The idea of primary and secondary rights follows the concept of transcendental and programmatic rights outlined by the German philosopher Wolfgang Kersting in his article on the political philosophy of international relations.

to set secondary rights absolute, such as political self-determination or the implementation of a desired political or social order – at the cost of ignoring and neglecting primary (transcendental) rights. In so doing terrorism not only turns the logical principle of the predominance of transcendental rights upside down, it also negates the predominant fundamental right to live, as this plays no role in terrorist rationales – neither in the sense of being a physical cause or motive to reflect its own maxim.

The upshot here is that the turning around of the logical principle of the preeminence of transcendental rights vis-à-vis programmatic rights implies in itself already dissolvable ethical aporias.

To this context of desired *consequence* – manifestation of a secondary right – and *incentive* – legitimacy of the claim – lacks the aspect of *means* and has to be added.

Given that the connection of desired result and intention is unavoidable and the use of forceful means therefore justified – which, as has been shown, appears to be problematic to vindicate to say the least – the next moral dilemma enters the picture, namely the likelihood that the amount of violence considered to be necessary will be exaggerated.

There is no doubt to the unbiased understanding that the moral legitimacy of the use of force is connected to another normative principle, namely the principle of proportionality. A principle which is twofold, on the one hand directly referring to the means used and on the other hand related to the one against which it is used.

In the latter respect we encounter anew the relation between perpetrator and victim – in Just War Theory represented by the combatant-noncombatant distinction – whereas in the former sense the reference is to the context of human and technological conditions (e.g. WMD).

Again, the asymmetric fighter neglects more or less all the ethical principles for the use of force - just cause, last resort, inevitability, proportionality, and discrimination in the immediate context of a perpetrator and victim relation. In addition, the ethical problematic is aggravated by the alleged inevitability of violence as a last resort for meaningful social interaction and the indiscriminate and disproportional use of violent means.

In short, what I wanted to demonstrate is that the justification of asymmetric violence – the use of violent means by terrorists or other sub-national actors devoid of any legitimate authorization – is facing almost insurmountable problems as far as ethical justification is concerned.

Existential Understanding of War

In this context we become aware that the instrumental understanding of war as Clausewitz most prominently expounded it transforms into a more existential one. (Münkler, 2002: 216-218) Though war and armed conflict remain ways of political expression, the difference results from the nature of the war-waging subject.

With the loss of the national and political monopoly over war and peace, armed struggle moves ever closer to its absolute form. The ideological war loses the moderating modes that normally limit warfare and armed confrontation. It is again Clausewitz who differentiates opposing yet at the same time mutually equilibrating principles in (symmetric) warfare. He discerns three escalating and moderating tendencies – mutual effects; reciprocal interactions – in war.

The escalating modes are

a) Both sides are driven to the extreme – Escalation if one side is committed to develop force relentlessly, the opponent has to follow through, which leads to the extreme;

b) No anticipation of opponent's intentions – Escalation results out of the impossibility to precisely anticipate an adversary's intentions (for it is relatively easy to assess powers and means, but more difficult to assess intentions);

c) Non-transparency of opponent's motives – Escalation as result of the impenetrability and non-transparency of opponents' motives and motivations.

However, when the warring parties acknowledge each other in their existential entitlements, those three escalating modes are balanced by moderating modes. These moderating modes are

a) Opponents normally know each other – and have thus basic insight into the other's customs and laws of war;

b) The time factor – in the sense that one side not undertaking any adversarial act induces the opponent to act accordingly;

c) The temporal validity of decisions – the temporary character of war decisions is based on the acknowledgement of the opponent as being a subject of international law and the idea that whatever is carried out within the commonly agreed rules can later be corrected politically.

This mutual acknowledgement between sovereign states by viewing the other as a *Justus Hostis* (just enemy) as outlined above, allowed for the instrumental waging of wars. This is to be found in political establishments only, where the domesticated character of war prevails. It is potentially no longer present in ideological wars where the enemy is no longer seen as an opponent, but rather as an adversary who has to be terminated. Wherever the claim is absolute and unconditional, implicit understanding and appeasement, which make moderation possible and keep war on an instrumental level, vanish.

Imperatives for the Future

To find answers to asymmetric violence will definitely be one of the cardinal challenges of the future – at least for the Western world. The competence of nation-states and associations of states to counter the arbitrariness of asymmetric forms of violence will be most crucial. Two aspects seem to be particularly essential.

First, apart from the cooperation of national and international institutions and organizations, a most decisive role will be played by the public discourse. If a terrorist shows no concerns about morality and moral legitimacy of his own acting, at least those who give him support should be concerned or withdraw support from him.

In modern information societies the mediation between political actors takes place via the media who, in their endeavor for sensations, become not scarcely the *"first accomplices of terrorists."* The media is further joined by all those societal exponents who in their mostly well-meant misunderstanding prop up what needs to be constrained - the claims of indifferent asymmetric violence.

Second, if indeed asymmetric Low Intensity Conflict in the form of terrorist and other non-state actors will accompany the future of established political orders, attempting to contradict their purpose and promise to maintain security and stability, a specific challenge is posed to regular law enforcement and military organizations. Their representatives – policemen and soldiers, who are obliged to abide by the rules and laws of warfare and legal and moral conduct – have to combat the inhuman (the immoral) without negating the human (the moral).

References

Amstutz, Mark R. (1999): International Ethics. Concepts, Theories, and Cases in Global Politics. Rowman & Littlefield Publishers, Oxford-New York.

Aristotle (1998): Nicomachean Ethics [translated by David Ross]. Oxford University Press, Oxford.

Stadler, Christian (2001): Internationales Recht - Völkerrecht und humanitäre Intervention, in: Edwin R. Micewski, et. al: Ethics and International Politics, Literas, Vienna, 24.

Christopher, Paul (1994): The Ethics of War and Peace: An Introduction to Legal and Moral Issues. Prentice Hall, Englewood Cliffs.

Chwaszcza, Christina; Kersting, Wolfgang (eds.) (1998): Politische Philosophie der internationalen Beziehungen. Suhrkamp, Frankfurt a. M..

Claudia Mongini (2003): Ursprünge der Asymmetrie in der Antike. In: Schröfl, J.; Pankratz, T. (eds.): Asymmetrische Kriegführung - ein neues Phänomen der Internationalen Politik?, Nomos, Baden-Baden, 15.

Clausewitz, Carl von (1976): On War [edited and translated by M. Howard and P. Paret], (Princeton University Press), New Jersey.

Karl von Österreich, Erzherzog (1974): Grundsätze der höheren Kriegskunst für die Generäle der österreichischen Armee. Biblio Verlag, Osnabrück, 1-2, zitiert in Andreas Stupka, Kriegsgeschichte und klassische kriegstheoretische Betrachtungen zur asymmetrischen Kriegführung. In: Schröfl, J.; Pankratz, T. (eds.): Asymmetrische Kriegführung. 41-56 (this passage translated from the German by E. R. Micewski).

Münkler, Herfried (2002): Über den Krieg. Velbrück Wissenschaft, Weilerswist, 91-116 and von Clausewitz, Carl: Vom Kriege. 216-218.

Kant, Immanuel (1993): Grounding of the Metaphysics of Morals [translated by James W. Ellington]. Hackett Publishing Company, Indianapolis, 36 [429] and 30 [421].

Lenckner, Theodor (1965): Der rechtfertigende Notstand. TübRA 14, Tübingen.

Tse-Tung, Mao (1966): Theorie des Guerillakrieges oder Strategie der Dritten Welt. Rowohlt Verlag, Reinbek, 45 (this passage translated from the German by E. R. Micewski).

Micewski, Edwin R. [et.al] (2001): Ethics and International Politics. Literas, Vienna.

Schopenhauer, Arthur (1995): On the Basis of Morality [transl. by E.F.J. Payne]. Berghahn Books, Oxford.

Schröfl, Josef; Pankratz, Thomas (eds.) (2003): Asymmetrische Kriegführung - ein neues Phänomen der Internationalen Politik? Nomos, Baden-Baden.

Sun Tsu: The Art of War (translated by Lionel Giles), Chapter Two: Waging Wars, 18.

Micewski, Edwin R. (2003): Moralphilosophische Überlegungen zur Legitimität von asymmetrischer Kriegführung. In: Schröfl, J.; Pankratz, T. (eds.): Asymmetrische Kriegführung - ein neues Phänomen der Internationalen Politik? Nomos, Baden-Baden, 31-40.

Kersting, Wolfgang (1998): Probleme der politischen Philosophie der internationalen Beziehungen. In: Chwaszcza, Christina; Kersting, Wolfgang (eds.): Politische Philosophie der internationalen Beziehungen. Suhrkamp, Frankfurt a. M., 36ff.

The Authors

ANNEN, Hubert, Dr. Studied psychology, pedagogy and neuropsychology at the University of Zurich. Senior lecturer for military psychology and military pedagogy at the Military Academy at the ETH Zurich since 1999. Head of the Assessment Center for professional officer candidates. Research in military pedagogy, evaluation, and validation of selection processes for military leaders as well as in personality and organizational attitudes. Holds the rank of major and serves as a militia staff officer in a territorial region.

BORN, Rainer, Dr. rer. nat., and Mag. phil. Studied at Oxford. Professor at the Johannes Kepler University in Linz, Austria. Teaches philosophy, theory of science and ethics and knowledge management to students in economics and computer Science. Research areas: cognitive science, model theory, semantics of science, artificial intelligence and communications theory. President of the Austrian Society of Philosophy from 1998 to 2000 and in 2002.

FLECK, Guenther, Ph.D., clinical and health psychology. Currently head of the Psychology and Educational Science Research Unit at the Institute for Human and Social Sciences, National Defense Academy, Vienna. Former assignments at the Austrian Military Hospital in the field of Aviation Psychology and at the Austrian Armed Forces Psychology Service as head of the Division for Psychological Training and Support. Guest lecturer in psychology of science at the Department of Philosophy of Science, University of Vienna. Vice-President of the Austrian Society of Organismic-Systemic Research and Theory. Research areas: political psychology, psychology of science and cognitive science, epistemology and philosophy of science. Related publications: *Self-Reflection and De-Reflection: Two Modes of Awareness Guiding Individual Experience and Action* (1999) and *Cartesian Anxiety and the Search for Certainty in Scientific Reasoning* (2000).

GUTKNECHT, Simon Peter, lic. phil. Studied psychology and business administration at the University of Berne. Research assistant in the Department for Military Psychology and Military Pedagogy at the Military Academy at the ETH Zurich since 2001. Research areas: personnel selection and evaluation, organizational psychology, job contention and retention. Currently working on a thesis entitled *Personality as a Predictor for Occupational Attitudes and Professional Commitment.*

KALLIOINEN, Outi, Ph. D. in education from the University of Tampere. Principal lecturer for knowledge management and coordinator for quality program for teaching and learning at Häme Polytechnic, University of Applied Sciences. Lecturer of military pedagogy at the Finnish National Defense College from 1997-2000. Recent articles: "The Development of Officers' Educational Expertise" (2002), "The Development of Teacher's Professional Growth" (2003); and "Ethics in Pedagogical Leadership" (2004). Master's thesis in Knowledge Management Program (2004, in Finnish).

KASTBERGER, Andreas. Major, Austrian Armed Forces. Since 2002, has worked at NCO Academy in Enns (Upper Austria) as an expert on military pedagogy and coordinator for civic education. Graduated from the Military Academy in 1990. Platoon leader and commander of a training company until 1998. Studied pedagogy and political science at the University of

Salzburg from 1993 to 1998. Subsequently a staff officer for military pedagogy and trainer for ethics and civic education at the headquarters of the Multinational Brigade South, Kosovo.

JUNG, Hermann, Ph.D. Retired from the Austrian Ministry of Defense in 2002. Head of the Defense Pedagogy subdivision in the Training and Education Section of the Austrian Ministry of Defense. Focus on military pedagogy and political education. Co-editor of the publication series *Studies on Military Pedagogy, Military Science, and Security Policy* and co-author of the book *Grundlagen der Militaerpaedagogik* (1996).

MATTSSON, Peter A. Dr. Senior lecturer at the Department of War Studies at the Swedish National Defence College. Received his Ph.D. from the Stockholm University in 1995. Doctoral thesis topic: *Generative Learning in Knowledge Intensive Industry*. Dr. Mattson's research focus: Theory and practice of military pedagogy; operational art and operational methods of leadership. Various publications on topics related to military pedagogy, military theory, and military leadership.

MICEWSKI, Edwin R. Social philosopher and political scientist, brigadier, and Director of the Institute for Human and Social Sciences (formerly Military Sociology and Pedagogy) of the Austrian National Defense Academy, Vienna. Member of the Science Commission of the Austrian Ministry of Defense since 1994. Visiting professor at the Department of National Security Affairs at the Naval Postgraduate School in Monterey, California. Research and teaching areas: political philosophy, the ethics of violence and force, military professionalism and leadership, and civil-military relations. Recent publications: *Ethics and International Politics* (2001); *Civil-military Aspects of Military Ethics* (2003); *Moralphilosophische Ueberlegungen zur Legitimitaet asymmetrischer Kriegfuehrung.*

OELKERS, Jürgen, Ph.D. in Educational Science from the University of Hamburg in 1975. Professor of general education at the University of Zurich and professor for education at the University of Berne. Research areas: history of education and the reform movements of the 19th and 20th century. Fourth edition of his book *Progressive Education* (1989) is due out by the end of 2004. Latest work is *Historisches Wörterbuch der Pädagogik* (co-authored with Dietrich Benner, two volumes, 2004).

PERRY, David, Ph.D. in Ethics from the University of Chicago Divinity School. Professor of Ethics at the U.S. Army War College in Carlisle, Pennsylvania since 2003. Teaches courses on ethics and warfare, strategic leadership, and U.S. defense planning. Previously taught ethics at Seattle University and Santa Clara University. Edits and writes material for web-based projects on just war theory. Publications: "*Repugnant Philosophy': Ethics, Espionage, and Covert Action,*" "*Just War Criteria and the War in Iraq,*" "*Our Accountability for Afghan Civilian Deaths: Some Insights from Shakespeare's Henry V,*" and "*The Problem of Total War in Judaism, Christianity and Islam.*"

RAVIV, Amira. Degree in psychology and Master's in organizational behavior from Tel Aviv University. Active duty lieutenant colonel, head of the Learning and Development Center of the Israeli Defense Forces Colleges, and head of the Military Psychologists Ethics Committee. Interests include organizational development and learning procedures of senior officers, counselling and implementation of strategic procedures, and promoting academic relationships with universities. Responsible for training and sociometric evaluation as well as for promoting staff teaching abilities.

ROYL, Wolfgang, Dr. phil., professor (retired) at the University of the German Bundeswehr, Munich. Dean for General Didactics at the Faculty of Education since 1976. Member of the German Association for Educational Sciences. Work includes publications on military education written in cooperation with the "Military and Social Sciences" in Strausberg, Germany; the German-Atlantic Association in Bonn; and the Hanns-Seidel-Foundation in Brussels.

TOISKALLIO, Jarmo, Ph. D. from the University of Turku in 1993 where he studied education, sociology, ethnology, and philosophy. Professor of Military Pedagogy at the Finnish National Defense College. Research area is the interrelations between educational theory, philosophy, and sociology. Currently establishing a networked research center for Action, Competence, Identity, and Ethics (ACIE). Recent publications include "Pedagogy of the Postmodernity" (2001, in Finnish) and "Being a Soldier in 2020?" in *Cultures, Values and Future Soldiers* (2002).

Studies for Military Pedagogy, Military Science & Security Policy

Editors:
Dr. H. Jung
Prof. Dr. W. Royl

Band 1 Edmund A. van Trotsenburg: Militärpädagogik. 1989.

Band 2 Hermann Jung / Heinz Florian: Grundlagen der Militärpädagogik. Eine Anleitung zu pädagogisch verantwortetem Handeln. 1994.

Band 3 Franz Kernic: Demokratie und Wehrsystem. Aufsätze zum Verhältnis von Gesellschaft, politischem System und Heer in Österreich. 1997.

Band 4 Edwin R. Micewski: Grenzen der Gewalt – Grenzen der Gewaltlosigkeit. Zur Begründung der Gewaltproblematik im Kontext philosophischer Ethik und politischer Philosophie. 1998.

Band 5 Rudolf Egger / Heinz Florian (Hrsg.): Pädagogische Professionalisierung im Bundesheer. Dokumentation und Reflexion des PädAk-Sonderstudienganges Wehrpädagogisches Management. 1999.

Band 6 Franz Kernic / Harald Haas: Warriors for Peace. A Sociological Study on the Austrian Experience of UN Peacekeeping. 1999.

Band 7 Franz Kernic / Jean M. Callaghan / Philippe Manigart: Public Opinion on European Security and Defense. A Survey of European Trends and Public Attitudes Toward CFSP and ESDP. 2002.

Band 8 Heinz Florian (ed.): Military Pedagogy – An International Survey. 2002.

Band 9 Edwin R. Micewski / Hubert Annen (eds.): Military Ethics in Professional Military Education – Revisited. 2005.

www.peterlang.de

Peter Lang · Europäischer Verlag der Wissenschaften

Heinz Florian (ed.)

Military Pedagogy – An International Survey

Frankfurtam Main, Berlin, Bern, Bruxelles, New York, Oxford, Wien, 2002.
275 pp., num. fig. and tab.
Studies for Military Pedagogy, Military Science & Security Policy.
Edited by Hermann Jung und Wolfgang Royl. Bd. 8
ISBN 3-631-39568-X / US-ISBN 0-8204-5988-7 · pb. € 51.50*

The book presents a broad, international discussion of military training and education problems. It uses the concept of Military Pedagogy to cover the full range of this discussion. As shown by the holistic articles at the beginning of the book, Military Pedagogy is viewed very differently by different authors depending on their national and individual backgrounds. Many of the articles use different subject matters to provide specific examples that, considered as a whole, show the great variety of tasks performed by Military Pedagogy. The two approaches, holistic on the one hand and detailed on the other, give the reader a comprehensive impression of what Military Pedagogy can be. Simultaneously, the specifically prescriptive articles provide many ideas for use by military training and education specialists. In this way, the book serves both theorists and practitioners. The international scope of the book is indicated by the list of authors who come from 10 different countries. The book also makes a significant contribution to improving the interoperability of Armed Forces.

Contents: Concepts of Military Pedagogy · Military Pedagogy in the changing world · Cognitive and mental models · From theory to practice in Military Pedagogy

www.ingramcontent.com/pod-product-compliance
Ingram Content Group UK Ltd.
Pitfield, Milton Keynes, MK11 3LW, UK
UKHW021014120926
8556IPUK00040B/1131

9 783631 528648